Anouchka Grose is a Bri
and author, who lectures
and Research. She is the a
non-fiction books. She writes for many publications, including the *Guardian*, and discusses psychoanalysis and current affairs on the radio, appearing on shows such as BBC Radio 4's *Woman's Hour*. Before training as a psychoanalyst, she studied fine art and was a guitarist and vocalist with the band Terry, Blair & Anouchka.

By the same author:

No More Silly Love Songs: A Realist's Guide to Romance
Are You Considering Therapy?
From Anxiety to Zoolander: Notes on Psychoanalysis

A Guide to Eco-Anxiety

How to Protect the Planet and Your Mental Health

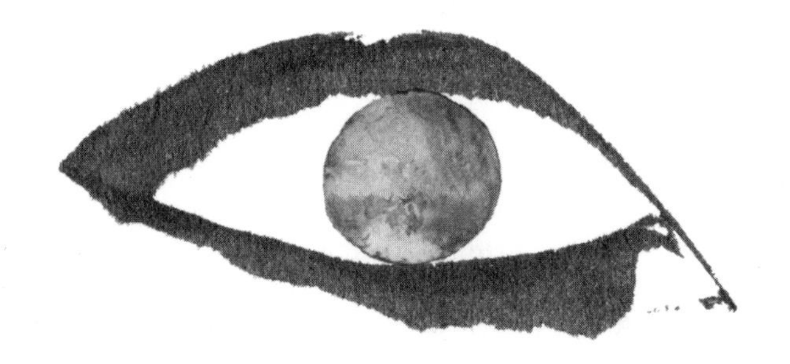

Anouchka Grose

This edition first published in the UK and USA in 2020 by
Watkins, an imprint of Watkins Media Limited

Unit 11, Shepperton House
89–93 Shepperton Road
London N1 3DF

enquiries@watkinspublishing.com

1 2 3 4 5 6 7 8 9 10

Typeset by JCS Publishing Services Ltd

Printed and bound in the United Kingdom

A CIP record for this book is available from the British Library.

ISBN: 978-1-78678-429-2 (Paperback)
ISBN: 978-1-78678-442-1 (ebook)

www.watkinspublishing.com

For Dot, who does so much to make a difference

"What is right to be done cannot be done too soon."

Jane Austen, *Emma*

CONTENTS

FOREWORD

Ed Gillespie

In his *Motto to the Svendborg Poems* published in the spring of 1939, Bertolt Brecht wrote: "In the dark times, will there also be singing? Yes, there will also be singing, about the dark times." As a climate activist for the last quarter of a century those words speak to me deeply. The grimly predictable "white swan" of the climate crisis has descended upon us, despite our best efforts, leaving many of us feeling totally hopeless. When the simple, linear, foreseeable trends in Australia – reduced rainfall, increased temperatures, drought, austerity cuts to forestry and fire services – combined through tipping points into complex, non-linear, chaotic and uncontrollable outcomes, in the form of raging "mega-fires", if you weren't getting at least a little bit anxious before that, afterwards you were certainly paying attention.

"Pyro-cumulonimbus" entered our lexicon, huge black fire-fuelled thunderclouds whose lightning strikes generate further fires, and we watched horrified as an area the size of Austria burned, killing half a billion animals, dozens of people and razing thousands of homes, lives and dreams to the ground. We perhaps began to realize then that nothing would ever quite be the same again. And then along came COVID-19 just to ram the message home a little harder.

As I write this, we are a month into lockdown in London. Two hundred thousand people have died globally, and hundreds of millions are housebound. Multiple strains of virus are on the loose, making a vaccine look elusive, and huge questions are hanging in the air, like the smoke after the Australian fires, around what sort of world we will inhabit on the other side of this crisis, what form of recovery we will need, and how to rebuild for what's next?

But what if this is not an isolated crisis? What if all these crises are actually inextricably linked? What if this is actually a *cascade of interconnected and interdependent crises* unfolding in both foreseeable and unforeseeable ways? A former colleague used to conduct a somewhat facile "thought experiment" with audiences, asking them "if they could wave a magic carbon wand to make climate change go away", would they? When some people demurred, she would cite this as environmentalists wanting "something to hate", as if there was something wrong with environmentalism. But what if the truth is that even if the magic carbon wand existed (spoiler alert: it doesn't), even if all the planet-stewing slew of emissions we have emitted in an explosive blink of geological time could be instantly removed, even if the oceans weren't acidifying dangerously enough to insidiously dissolve the calcareous homes of the marine creatures around them (less in your face than a 50-foot wall of forest fire racing towards you at 50 miles an hour, but ultimately no less deadly), even if that could all be dramatically, impossibly fixed in a moment … we would still possibly be fucked?

The truth is we'd continue to be chewing our way through the last wild places on the planet, and in doing so also chewing our way through bush-meat primates, undercooked bats or pangolins, in turn unleashing further waves of zoonosis, like the HIV/AIDS, Ebola, SARS, MERS and now COVID-19. We would still be facing the sixth mass extinction. We would still be destroying this unique (as far as we know) intergalactic Eden and the wider, wilder family of our shared natural heritage, and for what? To defend our right to consume? To dangle a cruel, false promise to the other six billion on the planet that the current lifestyles of the top billion are somehow achievable and replicable for all through "sustainability"?!

"Hang on!" I hear you thinking. "This is supposed to be a book about coping with eco-anxiety. Who's this guy? He's just making me *more* depressed!" And my first response to that is: *good.* Welcome to the difficult, horrible, painful, potentially fatal existential reality in which we find ourselves. And this is why

Anouchka Grose has written such a profoundly important and timely book to help guide us all through this darkest of times. We owe it to ourselves to be honest about the fact *this really might not end well* collectively, and individually the raw truth is that we're all going to die, at some point. No one here gets out alive, as Jim Morrison sang.

We humans are rather adept at the denial of our inevitable personal death. We numb ourselves with drugs and alcohol, pray for an after-life, or prop up our self-esteem through the contrived "immortality systems" of consumerism and shopping, as if life was a competition in which "he who dies with the most toys wins". We construct defences of identity for our own in-group's exceptionalism that drive the dangerous populism of our uniquely entitled "we", that allows us to "other" others to feel better about ourselves, and share pacifying myths excusing us of our own historical horrors of colonialism and slavery. Denial, distraction and disavowal ... we're pretty good on all these.

The cults of egotism, individualism and control, our arrogant hubristic heroism, have got us into this mess, and I believe it is going to take something very different, if not to get us entirely out of it, but to at least get us through it with the best of our humanity intact.

And let's be clear, *we are also lovely*. The empty streets, sports stadia and transport hubs in cities around the world right now speak to the biggest, most-selfless act of global solidarity we may ever see in our lifetimes. People are staying home to protect others: the weak, the vulnerable, the carers. We are connecting with neighbours we have never met before, rapidly building mutual aid networks. Few have questioned the enormous sacrifices being made economically by so many, because they can see the direct benefit on the lives of their fellow citizens. We're acting together, at speed and scale in shared purpose for the common good. It's impressive.

I call this the horror and the beauty. COVID-19 is almost a training run for the climate chaos to come; we're experiencing what Anouchka memorably introduces as a

Pre-Traumatic Stress Condition (it's not a disorder because it's actually quite a reasonable response to the science). And how we *react* to this anxiety matters. Will we fight each other in the supermarket aisles over toilet paper – probably not our finest hour – or find an unimaginable empathy for one another in the face of the uncertainty?

In *A Paradise Built in Hell*, Rebecca Solnit writes persuasively about how disaster-ravaged communities don't, despite the claims of lurid dystopians, tend to turn on each other in a Mad Max-style frenzy of cannibalistic tribalism. Perhaps this nightmarish vision is merely a projection onto the world by those traumatized individuals, usually authoritarians, of their own inner emotional conflicts? As Anaïs Nin put it, "We don't see the world as it is, we see it as we are." Instead, Solnit finds the grief of loss usually unites people through the common cause of their desperation.

Our challenge is to distinguish between different types of hope. There's the traumatized hope that derives from the denial that the world we have built might be fundamentally flawed. This is the hope that refuses to let go. It's the hope of environmental efficiency, sustainability and "one more push" environmentalism, which increasingly just doesn't stack up against the science of what needs to be done, let alone what we might actually be able to do within the so-called constraints of our commercial and political "realities". It's a hope that argues only positive optimism can carry the day, that fear is to be avoided and a suite of sizzling solutions will slide us smoothly into some techno-utopian futuristic fantasyland. It's a little narcissistic, speaks to the uniquely important "specialness" of people above nature and is, I suggest, afraid to really look itself deeply in the eye's mirror and explore its own shadows. It's the hope that actually perpetuates what is an abusive relationship between people and planet. That is not the hope we need.

I suspect we really need to *feel the fear* that comes from these anxious times. To understand the trauma of living, in what the ecologist Aldo Leopold described as "a world of wounds", and perhaps most importantly not to *deny* the grief of the incalculable losses we have already inflicted and are inflicting, but to *embrace* it. *And then to change.*

As a young marine biologist I remember being moved to tears at the decimation of our Blue Planet's ocean riches by indiscriminate industrial fishing. When we've cut down half the world's irreplaceable, on any vaguely human timescale, primal forests, killed half of all wild vertebrates in my own lifetime, and are facing a hotter, more turbulent and unforgiving world climatically and it seems pathogenically, it would surely be wrong not to feel this as anxiety and grief? Is the alternative not delusional? Or the equivalent of a pacifying pill?

That's not depressing fatalism, "doomsdaying", welcoming the apocalypse or all the other dismissive ways in which traumatized hopeful optimists disparage those who choose to grieve. It's realism. "Apocalypse" is actually about drawing back the veil, not Armageddon, so we can see the world as it really is. It's revealing our love and connection to that which we have lost and are losing. *We grieve because we love.* And grief is not the paralysis of despair, it is the dynamic process, as Anouchka notes, that takes you to a new place.

Which brings us to the hope we really need. The grounded hope on the other side of grief. I think one of the reasons we hopefully get at least a little wiser as we get older is because we have encountered, experienced and lived through heart-breaking loss. Brushes with mortality remind us of what really matters. Appreciating the inevitability of our own demise, and the prospect of also losing most of those we love who will go before us, does not make us love our lives less or live them any less fully. Quite the opposite in fact.

In the space of the last four years I lost my dear father, became one myself, and then my darling middle brother died suddenly and tragically. Pretty much the horror and beauty of the cycle of life summed up there in 42 short months. As I waded through the deep waters of loss, the irregular tides of grief washing over me, often erratically and unexpectedly, what

I learned was that it was the very ephemerality of life that made me love it all the more. The flower is not less beautiful because of its composted ending, it is *more so*. And its bloom counts. Especially if you're a passing bee.

My heartbreak taught me that you don't "get over" the grief of loss. You get *through* it. Beyond the Kübler-Ross five stages of grief, which are more like looped and backtracked intermingled phases in my experience, we overcome our denial and anger, drift through the depths of depression, stop trying to make impossible deals with the divine and ultimately transcend to acceptance.

The counsel of wisdom is not for the pain to be removed or solved, in the same way therapy is not to fix or normalize people. It is totally OK to be really, *really* upset, think difficult thoughts, succumb to catastrophizing feelings and generally lose your shit a bit. Perhaps this is all part of the unavoidable pain of the human condition? Ultimately it is about how we sit with this discomfort. How we ask for help.

We are living through an unimaginable future today. Each morning seems to bring another previously inconceivable thought experiment into reality. What if the world burned and we couldn't stop it? What if we effectively locked a billion people indoors for several months? What if we grounded virtually every plane in Europe? What if we had a negative oil price and companies were paying people to take it away? What if the machine stopped? What if *we* stopped?

As birdsong rings around our cities where it was previously drowned out and choked by the traffic, as the Himalayas loom out of the dissipating smog over northern India for the first time in three decades, as deer, goats and wild boar prowl urban suburbs and the canals in Venice run clear as glass, we are witness to the reality of what we have long been told is an *impossibility*. A rapid rewilding as civilization holds its breath for fear of inhaling the unwanted. These are the days the Earth stood still.

The fake Venetian dolphins that spread virally around the world like-you-know-what I think did so because they spoke to something very deep and vital inside us, a longing if you will, for what Anouchka calls the "charisma of birds", others "biophilia" or most of us the visceral, embodied love of nature and the wild. Unsurprising, as we are of course an inextricable part of that restless, teeming, seething, trembling congregation we call "life", as this crowning phage is so ruthlessly reminding us.

This is something we have denied ourselves since the Enlightenment, and the process of our separation from this living family, our arrogant elevation above it, our selfish exploitation of its wealth as utilitarian resources not comrades, entities or partners, let alone the horrendous things we have done to each other, have I believe left us psychologically scarred and, yes, traumatized.

But this is also perhaps where the sinuous answers to our modern malaise wriggle back, radical and root-like, into our genuinely grounded, emerging consciousness. If you awaken to the notion that the trillions of cells that make up the body we might arbitrarily call "you" are barely half "yours", made up as they are of bacteria, fungi, microbes and of course viruses. If you can find the awe and wonder in that magical interconnectivity, that you are, as philosopher Alan Watts used to say, "not born into this world, but out of it, like a wave from an ocean". If you can feel the strange beauty of your own personal insignificance in this collective magnificence, then that very baring of your soul to the world may just begin the essential journey of reconnection, reconciliation and resolution.

Almost every supposedly "primitive" indigenous culture on the planet has known this. Initiations and rites of passage exist specifically to take people, often the most potentially problematic hot-headed young men, through difficult transformative experiences that dismantle raging aspects of their fevered egos. Right now, we are being subjected to a planetary-scale initiation. A potential transformation. It is dangerous, worrying and most certainly uncomfortable. But that's how initiations are meant to be.

This is perhaps a necessary humbling if we are to get beyond the superficial, self-help "McMindfulness" of so much that passes for psychological support these days. The practices of wild generosity and radical friendliness that Anouchka prescribes as the antidotes to our individualistic malaise would be familiar to our ancient ancestors. And those are the behavioural responses to the adrenaline- or cortisol-addled amygdala moments we're all experiencing that we really need.

It's right to feel anxious. But maybe even more than our volatile climate, the lion we really fear is ourselves. As East German dissident Rudoplh Bahro said, "When an old culture is dying, the new culture is created by those people who are not afraid to be insecure."

We're not going to "beat" climate change. It's not a binary, polarized win-or-lose scenario. We are going to hopefully live through it. And in doing so we will be changed by it. We will be a gentler, kinder and more thoughtful culture. I think we'll have to be. But like any effective therapeutic transformation, this starts with acknowledging we have a problem, which turns our anxiety from unease to *eagerness to do something*, and the requisite willingness to act. And that's when the real work begins. Like all transitions it will not be painless. But this urgent and impressive book is an essential guide as we embark on this journey together.

If we can sit and be still with our discomfort, celebrate the awe and beauty of the nature we still have, alongside the friends with whom we share it, and be appreciative and grateful for all these gifts, then even in these dark times we can sing together.

And that for me is where the radical active hope lies.

Ed Gillespie is a Director of Greenpeace, UK and a Trustee of Energy Revolution.

He is a keynote speaker, serial entrepreneur and futurist, and the author of Only Planet – a flight-free adventure around the world. *You can follow Ed on Twitter @frucool*

INTRODUCTION

I'm writing this at the end of the first week of the UK's official COVID-19 lockdown. The world outside seems to be changing beyond recognition. I go out once a day to walk the dog and the streets are all but empty. It's the beginning of spring. Cherry trees blossom and wilt without their usual audience. Strangers cross the road to avoid one another. Last night I attended my first Zoom dinner party. There have been just under 31,000 deaths worldwide so far and we hear that the crisis in the UK is just about to tip over into barely manageable levels. I have no idea what kind of world we will be living in by June 2020, when this book is due to be published – let alone by autumn, Christmas, next year. Already all films and TV shows look a bit weird: crowds, cheek-kissing, bars and restaurants. So quaint! Where are the facemasks? Things whose existence we took for granted, even while recognizing their inherent dodginess – flights, fast-fashion, the high street – might barely exist once we emerge from this for all I know. Or maybe, we'll experience the ultimate gaslighting from governments and corporations and we'll be fast-tracked back to capitalist business as usual, no lessons heeded.

This is a book about the apocalypse or, more particularly, about the apocalypse we feared before it started to happen. People who suffer from eco-anxiety are acutely aware of how fragile our systems are, both man-made and natural. This horrifying knowledge is what keeps us up at night. The COVID-19 crisis has already revealed how rapidly *everything* can change. Schools can close, exams can be cancelled, jobs can be lost, businesses can buckle, right-wing governments can switch to socialist principles overnight. (And then, maybe, they can flip back again with equal alacrity.) Those of us who are attuned to the looming climate disaster have been dreading all this stuff for years – pandemics are very much on our list of things to worry about. Global heating and other environmental disturbances such as humans interacting with wildlife due to trade or habitat loss on a scale that was never meant to be, could facilitate the development of more novel viruses such as COVID-19, and the transmission of infectious diseases is likely to be increased in ferocity due to the extreme weather events of a hotter planet.

So how useful is it to have a book about managing our sense of dread, now that we find ourselves in something very much in the guise of the future we were desperate to avoid? My hope is that our anxieties will at least have primed us somehow – we haven't exactly been taken by surprise. And on the bright side, if the book is about managing our fears around massive change, we're definitely about to be given plenty of chances to test the theory.

Our individual lives are played out in the context of the wider world, and our innermost thoughts are interwoven with perceptions that arrive from outside. If the very planet we live on seems to be on course for total ecological, socio-cultural, economic meltdown, how can any of us hope to keep going? Why would we get up, go to work, study or do anything short of clinging to a tree and screaming? Whereas "talking about the weather" used to be synonymous with having a boring chat, nowadays it's just as liable to incite panic. Like many of my colleagues, my friends, my family and the people who come and talk to me in my therapy practice, I'm anxious about climate change. In fact, I find thinking about it almost unbearable. If people come to my office and describe staying awake all night worrying about burning forests and melting ice caps, I'm not going to tell them to stop. They're right: it's the people who *aren't* worried who are crazy.

The problem is how to respond to these kinds of anxieties in a way that's actually helpful – both to the person, and to the environment. For instance, perhaps a responsible therapist shouldn't be trying to offer easy relief. Maybe a person's anxiety can be put to good use, doesn't need to be immediately dismantled, and is actually an adequate and sane response to a dire situation. It's possible that a certain level of psychological distress can be part of the process of dealing with things. As Greta Thunberg has said, "I don't want your hope, I want you to panic." Maybe we can't, or shouldn't, expect to feel happy, safe and well all the time, because that's just delusional: a fantasy fed by the irresponsible socio-economic ideologies that got us into this whole mess in the first place. There's no rational reason why humans should expect to live our whole lives like pampered pets. Still, staying awake, tearing your hair out, is unlikely to make you an effective activist, let alone a regular, (mostly) functional person. There has to be a position between climate-related nervous breakdown and ostrich-like denial: a serviceable state from which you can think and make changes, communicate and act.

Climate disaster, and its accompanying social upheaval, is a problem that none of us will be able to deal with on our own. Talking about the reality of the changes that we face, according to certain studies, can be helpful. And staying informed, although often alarming, is a vital part of constructive engagement. Storms – both meteorological and political – are brewing, and we might as well prepare to weather them as graciously, intelligently and humanely as our uncertain futures will allow.

This is intended as a book that helps you think differently about the escalating crisis, and offers ways to respond to it thoughtfully. My own carbon footprint is far from neutral, but I'm as worried as the next person – to the point where I don't own a car, I only buy second-hand clothes, and I've been mostly vegetarian since 1985 (apart from a dreadful long-term lapse while I was married to a meat-loving Argentinian). I also hear a great deal about other people's climate anxiety in my practice as a psychotherapist, as well as sometimes being lucky enough to hear about the ideas that help them to stay above water (literally). So, while I can't promise to be a beacon of climate virtue, I can promise to want to understand all this stuff better, and to present some information that might help you.

I can also promise to be open to any intriguing ideas, however idiosyncratic. In my opinion, therapy should emphatically *not* be used to normalize people, nor to support the status quo. Therapists are often able to help people far more by being prepared to put norms into question, rather than by cajoling their clients into fitting in better. So I believe we need to be thinking about critical cures – not about quick symptom relief, and not about placating people either by downplaying the problem or by blotting it out with medication. Any serious response to climate anxiety has to set out from the premise that *ecological breakdown is real*, and the people who are worried about it should have their thoughts and fears taken seriously. And of course, not everyone has the luxury of being in therapy; we're all going to need to be there for each other in whatever ways we can, listening, understanding and taking action when necessary.

As we speak, research is being done on the ways in which humans have dealt with other kinds of overwhelming change; for instance, how American Indians have dealt with being ripped out of the environments they and their ancestors grew up in, and relocated to reservations. What could those people do to help themselves tolerate their new circumstances? You can fight to the death, but you might be the loser. Or you can find ways to convert your shock, rage and helplessness into some kind of transformative experience, enabling you to think about life on completely new terms. So, alongside all the work that needs to be done on changing economic systems, changing consumer behaviours, changing laws and changing farming practices, we also need to work on learning how to manage, and even harness, our feelings *and, above all, to learn to tolerate uncertainty*. The kinds of conversation that seem most likely to be able to meet that challenge are those that open up radical questions about existence and coexistence, and that definitely *don't* discourage people from thinking difficult thoughts.

If you suffer from eco-anxiety, this book alone won't "cure" you. It may be that it would be helpful for you to talk about it in therapy, join a group with other sufferers, practise meditation, get more involved in activism, or all of the above. But I hope that I can at least point you in helpful directions, and make you feel less alone with the problem.

Some of the suggestions in this book might sound a bit incongruous coming from a psychoanalyst – we have a reputation (deservedly) for being dubious about human "goodness". However, above and beyond my Freudian leanings, I'm a *huge* Jane Austen fan and actually believe she might have more to offer the climate movement than a lugubrious Austrian pessimist such as Freud. While both thinkers are engaged with feelings, change and self-improvement, I think Jane has a particularly astute take on being a good person for other people, while not making yourself miserable in the process. If this book has a bottom line, it's that. Call me naive, but I actually believe it's possible to *decide* to be a better person – if Emma Woodhouse can do it, we all can. So I apologize in advance if I get into an overly Freudian, morally ambiguous tangle here and there. I promise I will eventually come out the other side with a sincere and simple message: let's all try to be really, really nice, even if we have to suffer a bit in the process. (Great, now I've frightened all my colleagues away, we can have a proper talk …)

Being upset is OK. Being upset is actually part of the solution. *Anxious people unite!* Denial, distraction and disavowal are the problems – although a certain level of "functional denial" is undoubtedly necessary – whereas anxiety, unhappiness and even anger are all states that can work against complacency; in themselves, they seem to demand change. So be anxious, be very anxious, because your anxiety can be a brilliant resource. And I hope this book can help you – and me – to see how …

1

ANXIETY: FREAKING THE F**K OUT

> *"I'm no longer accepting the things I cannot change. I'm changing the things I cannot accept."*
>
> Angela Davis, philosopher and political activist

I don't mean to make things any worse but the Amazon is on fire, Greenland is melting, the Maldives are drowning, the Great Barrier Reef is going grey, the very air we breathe is on the verge of turning against us … and we're currently in the middle of a global pandemic, with the promise of more to come. What's not to feel anxious about? It might appear self-evident that anxiety is the only sensible response to the climate crisis. More than that, it can sometimes seem that not being anxious, even for a minute, is a kind of thought crime against the planet. How dare you stop worrying and start feeling OK! Don't you know how serious things are? Lose concentration for a second and you risk waking up to find yourself eating a beefburger on an aeroplane.

Still, if you're going to stay functional enough to keep existing – let alone appreciating the miracle that is life on earth – you will need to get some rest at some point. But anxiety isn't something you can switch on and off at will. It can take you by surprise when you're least expecting it, and it can stay with you no matter how hard you try to reason your way out of it. Anxiety about the future of the planet emphatically doesn't fall into the category of "Don't fret about things beyond your control." The problem with climate-related worries is that you are liable to feel both compelled to act on them – even to give your entire life over to acting on them – and at the same time feel utterly helpless, in the knowledge that your actions alone will make very little difference. It would take millions of people to act, drastically – but more importantly, a small yet significant number of politicians and heads of corporations – and so far, so bad on that front. The frustrating thing for climate activists has been to see just how quickly change can be effected when a crisis is taken seriously. All the immediate actions taken to prevent the spread of COVID-19 have had an immediately beneficial impact on the environment: flights grounded, reducing traffic on the road, evaluating what "essential business" truly is. This has highlighted that there is scope for these changes to be made: it's *not impossible*. You can't afford to give in to helplessness at the same time as you can't actually solve the problem. Or certainly not all by yourself. It's enough to drive a person mad.

In this chapter, we'll take a look at anxiety from a number of different angles to try to see what it is, why we feel it, and what we might be able to do about it. In the chapters that follow we'll look at some of the ways we can channel our anxieties into activities that are actually helpful. Ultimately, the point won't necessarily be to stop being anxious altogether, but to learn to be on better terms with our worries so we can use them to orient us, energize us and maybe even bring other people on board. As they say, a problem shared is a problem halved, and this couldn't be truer than in our current ecological situation. If, instead of a totally freaked-out minority, we had a realistically concerned majority, then we'd all stand a much better chance. At the same time as bringing our own anxieties down to more habitable levels, we might also want to think about raising other people's up to less-lackadaisical ones (more on this in Chapter 4).

What is Anxiety?

"Anxiety" is an interesting word because it has two subtly different meanings. For some reason the second one seems to get forgotten, or drowned out, by the noisier first one. When psychiatrists talk about "anxiety disorders", they're working with the first definition: a feeling of unease generated by situations with worrying outcomes. This becomes a disorder when the anxiety gets so intense that the person is no longer able to do the things they need to do to stay on top of their own life. Hence it can sometimes seem that anxiety is necessarily "bad". Some definitions of anxiety even begin by seeing it as something pathological in itself – the *unhealthy* counterpart of fear. If fear is a reaction to a clear threat, then anxiety is a reaction to an *unclear* one: in the case of fear there is a tiger, in the case of anxiety there *might* be one. Therefore, in the second case, you could have saved yourself some bother by having a more optimistic outlook; more fool you. Anyhow, anxiety is unpleasant for the person experiencing it, and the first definition of this state contains the possibility that the bad thing isn't actually going to happen. This uncertainty can, of course, be worse than the sure knowledge that you're somehow done for. At least when you know it's game over, you can simply resign yourself …

So, if you say you're suffering from anxiety rather than fear, you're already inviting the suggestion that you're "just being neurotic". But in the way that "climate change" has been upgraded to "climate crisis" in the more ecologically responsible news media, you could perhaps say that "climate anxiety" should probably be changed to "climate fear" to make it very clear whose side you're on when it comes to gauging the reality of the threat. Still, climate anxiety, or eco-anxiety, is the phrase that's taken hold, so let's work with it. And the surprise advantage of this might be in the second meaning of the word "anxiety", given on the Cambridge English Dictionary website as "eagerness to do something". Examples of this second definition might be "Ben was anxious to give his sister the perfect birthday present" or "I was anxious to be on time to meet you." Although it might still contain a hint of discomfort, the emphasis is on being *willing to act*. While the first definition of anxiety suggests a kind of frozen passivity, definition two is more energetic – more positive.

While you wouldn't wish an anxiety disorder on anyone, we might be able to think about different gradations of anxiety, and to see that there are perhaps better and worse ways to be anxious. With the first definition you risk being unable to act, while with the second you are more likely to be spurred into action. So how can we help ourselves to be the "right" kind of anxious? Perhaps, as a starting point, it could be helpful to look a bit more carefully at how anxiety works on the body and mind.

Anxiety: Not All Bad

Anxiety could be said to be one of the many things that's made humans so well equipped to survive and proliferate. It compels us to think ahead and to plan for things that might go wrong. You could say it's something like fear plus language and a sense of sequential time. In physiological terms, it puts our bodies into emergency mode. Obviously, it's great that we have this system, because life provides all sorts of emergencies; anxiety is one of our many spectacular adaptations. We're part of a food chain which, in the past, meant we were liable to be eaten by bigger, fiercer animals. As is the case for all animals, natural selection has meant that we have adapted to our environment, although, also like all other animals, our circumstances might alter, meaning that what was once a life-saving bonus begins to count against us. (We're also *unlike* other animals in that we have technology and global organizational systems, but more on that later ...)

For instance, you might become the biggest animal in the forest, able to eat all the other animals. What happens once you've gone and done that? This is basically the human situation at the moment – our extraordinary adaptations are shooting us (alongside other lifeforms) in the foot. But the thing that used to be good about anxiety is that it prepared us to deal with threats. So perhaps it can help us out with this one, if we're smart about it.

Chemically, our fear response and our anxiety response are pretty much the same. Our sensory system – whether through looking, listening or reading the newspapers – tells us that something is threatening our safety. Our amygdala, a part of our brain that handles emotional processing, alerts the hypothalamus to the possibility of danger, exactly like that of the zebra who suddenly looks up, sensing a lion. While the brain's neocortex simultaneously goes to work on assessing the level and nature of threat, the hypothalamus simply gets the body ready to deal with whatever it might be – it doesn't even need to know that much, it just produces a generic response to "bad things". It does this by triggering the release of the so-called "stress chemicals" adrenaline and cortisol. This chain of bodily events is known as the sympathetic nervous system. Adrenaline has speedy, and extraordinary, effects on the body. It causes the heart to beat faster, thereby pumping oxygen to the brain in order to help it think. In this state, not only can we put ideas together more efficiently, but we can even hear, feel and smell better. At the same time, it also causes our breathing to speed up, and extra passageways to open in the lungs to help us to process the increase in oxygen, feeding it into the bloodstream so the heart can hurry it along, and so on.

As well as all this, some of our fat and glucose supplies are mobilized, giving us an extra shot of energy – in case we need to fight or run – and our immune and reproductive systems are suppressed because we won't be needing any of that for now, thank you. The hormone cortisol is released in a second wave, so long as the perceived threat is still operative. It keeps the body on high alert to stop us becoming complacent.

In time, ideally, the danger will be warded off and the other half of the system – the parasympathetic nervous system –

will kick into action, soothing things back to normal again. Once your mind has persuaded itself that you are no longer in danger, your brain starts sending messages out to calm your body down. Your breathing goes back to normal; your heart slows down. But problems come when there's no reason for your brain to tell your parasympathetic system to get going. What happens if the threat doesn't go away? This may be the case whether it's a bullying boss whom you have to see every day at work, or the looming threat of societal collapse due to climate breakdown.

There's also the fact that different people's nervous systems operate differently, some being very good at triggering appropriately and then calming down again afterwards, while others might kick into action at the drop of a hat (sometimes literally, say in the case of a person with germ-related OCD – I used to go out with someone who would genuinely have a meltdown if ever they dropped their hat on the floor). The problem with never releasing your body from its state of emergency is that this can result in serious organic damage. Being constantly flooded with stress chemicals can eventually lead to high blood pressure, heart problems, lower bone density, decreased cognitive function, immune problems, reduced fertility, strokes, digestive issues, blood sugar crashes leading to mood problems and exhaustion, and abnormal appetite and fat storage (especially around the abdomen). This is because your body is *not designed to live in a constant state of fear.*

How Eco-Anxiety Affects Your Mood

In the "fight-or-flight" system, one of the bodily acts that is most dispensable is the ability to feel joy, or to laugh. Your nervous system quite understandably deems it unnecessary to be able to feel carefree when you're running away from a tiger. So if your daily anxiety is triggered by seeing people merrily using disposable coffee cups, or idling engines contributing needlessly to unbreathable air, you are less likely to able to feel light-hearted and joyful. Your brain locks out creative thought and imagination when frozen within the sympathetic nervous system, and arguably, creativity and imagination are exactly what are needed for us to find ways out of this global predicament. Therefore, it's really important to try to find out what works best for you when it comes to reducing anxiety, and I hope that by the end of this book you'll have a clearer idea. Everyone's bodies and minds work differently, so it's not a matter of being told what to do, and then following it to the letter. It's far more useful, and realistic, to have a good working knowledge of the options – some of which might sound as silly as re-watching old episodes of *Gilmore Girls* – and to do the things that work best for you, rather than the things that sound "serious" or virtuous.

> "We visited friends at the weekend and they got takeaway coffees at the park with our kids. I didn't get one: I'd forgotten my keepcup and my rule is: no cup, no coffee. I asked in a non-judgemental, friendly way, 'Do you get takeaway coffees every day?' He gets one every single morning, sometimes one the afternoon too, always in a throwaway cup. He said that taking a reusable cup out with him 'wouldn't be very convenient'. I often think of his comment, and it makes me feel sad and alarmed, imagining all the millions of people out there, just like him, just one throwaway plastic cup, millions of times over, every day …"
>
> Annie, writer

Anxiety as a Chemical Reaction

People can be quite fatalistic about the functioning of their bodies and minds. You either *are* something, or you're not. Popular neuroscience can make us feel entirely at the mercy of the arbitrary chemicals in our own flesh. If you believe you just happen to be a person whose serotonin levels are low, or whose parasympathetic system is lacklustre, you might believe you just have to suck it up, or to look for a chemical cure. What this strand of thinking fails to address is that *ideas* can have a huge impact on the functioning of the body. This isn't New Age mumbo jumbo, but scientific fact. More than that, it's just blatantly, demonstrably obvious.

To put it simply, if you *believe* there's a burglar in your house you are likely to have a panic reaction. You will feel all of the effects outlined above – your heart will race and your mind will go into overdrive. But if your sleeping partner convinces you that it's just the neighbour's cat, who has knocked over the plant pots by your back door again, then your body may auto-correct without too much trouble. The burglar and the cat are simply ideas – you haven't actually witnessed either of them – but the chemical effects on your body and mind are real. Equally, if your cat-owning neighbour tells you that your sleeping partner is cheating on you, you might have an intense emotional/physical reaction. It could turn out to be the case that your neighbour is mistaken – so your physical reaction was triggered by their erroneous idea. Ideas, thoughts and images certainly have effects on the body, which can be unfortunate if you have a pesky neighbour, but may be helpful if you can track down some ideas that are robust enough to help you through difficult experiences. While people's bodies might be very different, it probably isn't wise to be too fatalistic. Ideas, not to mention real events, can have a massive effect on your wellbeing – in either direction. Actions and ideas can make you ill – but may also help to make you better.

To sum up what we have found so far: there's anxiety and its effects on the body, and then there's the question of how we might be able to respond appropriately to that in cases, such as climate anxiety, where there's little chance that the threat will simply go away.

Responding to the Body with the Body

The first approach – perhaps most readily recommended in mainstream, contemporary medical circles – is to deal with bodily effects using bodily means. This can either mean taking medication or otherwise "tricking" the body out of its uncomfortable state, thereby bringing the mind back into line as a knock-on effect. The latter approach might include breathing techniques, meditation/mindfulness or exercise – all ancient techniques thoroughly approved by the contemporary medical community.

How medication works to alleviate anxiety

Psychiatric drugs work on the brain in a number of ways. The most commonly prescribed category of drug in cases of anxiety are SSRIs (selective serotonin reuptake inhibitors) such as citalopram (Celexa) or fluoxetine (Prozac). These affect the brain by keeping the neurotransmitter serotonin held up on the pathway between neurons (in a space known as the "synaptic cleft") rather than being reabsorbed on arrival. These drugs are "selective" in that they mostly leave the other neurotransmitters uninterrupted, focusing their attention only on serotonin, the so-called "happiness hormone". On finding itself stuck in this liminal space, poor old serotonin keeps attempting to make its way through, like a person on the way to the buffet carriage of a train who has unfortunately become stuck between two carriages. The effect of this is that the neuron apparently keeps receiving the message "happiness is appropriate right now" as if it's a new arrival of serotonin each time – and the human host of the neuron thereby becomes much more cheerful. (Of course, it's more complicated than that, as serotonin also has an impact on other things, such as appetite, social behaviour and aggression; hence SSRIs aren't *always* effective as a mood stabilizer.)

In a similar category, there are SNRIs (serotonin-norepinephrine reuptake inhibitor), which also inhibit the reuptake of noradrenaline, the key neurotransmitter in the sympathetic nervous system – the "fight-or-flight" chemical. This might seem a slightly less-intuitive treatment for anxiety – why keep a stress chemical in circulation for longer? – but all these newer drugs have surprising and not always predictable results, which can either work for or against you. Some people find they really help, while others find they don't. Most doctors would only try this second option once they'd exhausted the first.

After that, you have tricyclic antidepressants, an older type of medication that's been largely overtaken by SSRIs. These increase the levels of serotonin and norepinephrine, while blocking the effects of acetylcholine. While this may sound like quite a similar approach to the previous one, it seems that this category of drugs has more unsettling side-effects than the former, such as blurred vision, racing heart, dizziness, plus digestive issues and skin trouble.

Lastly, there is the option of taking sedatives (also known as benzodiazepines) such as Valium or Xanax. These are sometimes given to people who turn up at their doctor's surgery in a terrible state, or are taken alongside SSRIs for the first couple of weeks or so, while the long-term treatment begins to become effective. These drugs cause drowsiness and can make you feel pretty weird, as well as being addictive. Benzodiazepines work on the brain completely differently: their effects come from increasing the effects of gamma-aminobutyric acid (or GABA). In this sense, they affect us more like alcohol or opiates do, impacting on our abilities to rationalize as well as to perform physical actions. GABA is an important neurotransmitter in the parasympathetic nervous system – it helps to calm us – but the long-term use of "benzos" can lead to depression, disordered thought, poor judgement and problems with motor coordination (which may result in physical accidents such as falls or car crashes), sexual dysfunction, constipation and trouble sleeping. Doctors generally don't let people take it for more than a month at a time, making it an unlikely "cure" for eco-anxiety, which is most likely to be something experienced long-term.

Medications are certainly available to people suffering from climate-related anxiety, and they may be helpful either in the short or the long term. You could be cynical and say, "Why should people have their anxieties switched off when they're actually about real things? Isn't that a bit like putting them in the Matrix?" While that's not a completely invalid point, it's also true that medication can give people a bit of respite and, who knows, might even bring them back to their everyday lives with a vengeance so they can be better activists for the cause. It's not the case that modern medications simply subdue people and make them more biddable (like in the movie *One Flew Over the Cuckoo's Nest*). If you think you could be helped by SSRIs, for example, there's no reason not to try them.

Arguably, it would be an unusual doctor who medicated someone for their climate worries without also suggesting other non-chemical options. These could be anything from Cognitive Behavioural Therapy (CBT), to Mindfulness-Based Cognitive Therapy (MBCT), to running, swimming and yoga. So before talking about therapy, let's delve deeper into these techniques, and have an overview of *why* exercise, deep breathing and meditation can be helpful.

Exercise as medication

Aerobic exercise such as running, cycling, dancing, and even walking, has effects on all the brain chemicals mentioned above. Exercise can bring down cortisol and adrenaline levels by *completing* your body's fight-or-flight response, although sometimes it increases them first. While your body might call on some adrenaline to get itself through a burst of activity, regular exercise tends to reduce our resting and sleeping levels of stress hormones. The exception to this is over-exercising or training for competitive sports, both of which can be very stressful. When doctors recommend exercise as an antidote to anxiety, they're suggesting you do it in a non-anxious way. Walking or cycling to the shops and back is good; trying to break the four-minute mile by the end of the month, not so much. Alongside bringing down stress hormones, exercise also gives you a shot of endorphins, which work like opioids on the brain. These are responsible for what's known as "runner's high" – that feeling of euphoria you might get after a bout of exertion.

On top of all these excellent drug-like effects, exercise can also be mentally distracting, not to mention socially lively – both of which are also great things with fantastic neurobiological effects. *Yay! woo!* etc. So when a doctor or your mum tells you that exercise is a good idea and will help you to feel more optimistic, they're not always necessarily simply trying to fob you off and make you responsible for your own unhappiness; "get some exercise" can often sound like an insult to people who are suffering from deep existential angst. On the bright side, this advice is underpinned by testable scientific thinking. But it can also sound like a way of saying, "I don't want to listen to your endless moaning." Or even, "Stop fretting about nutty stuff like the chemical make-up of the atmosphere and focus on your glutes." While positive research findings on the mental effects of exercise are a gift to politicians who want to withdraw funding from organizations such as the NHS, and to insurers who don't want to offer pay-outs (telling people to exercise is free, and it's them, not you, who looks bad if they don't then go and do it), it's also a really good idea to get some exercise.

"My dog is saving my life right now. I have to go out and walk her in green space, every day. I have to 'be in the moment', with her, meet her needs for play and fun. She's living richly and fully and has no concept of future loss, and that rubs off on me when I most need it. And it also means I'm moving my body which lets me let go of tension or meander over complicated feelings outdoors which I find really helps. I feel like everyone with eco-anxiety would benefit from getting a dog!"

Antonia, doula

Meditation as medication

At the other end of the physical spectrum, we have bodily techniques such as deep breathing and meditation that slow you right down. Although human beings have been using these techniques for millennia, it's only recently that expensive, sophisticated machinery has been added in to demonstrate that people were on the right track in the first place and should just keep doing what they were already doing. A new treatment area known as "biofeedback" has been made possible by encephalograms, a.k.a. EEGs (which record brain waves), and wearable sensors that show us how the body responds to a variety of stimuli. In a biofeedback session, you will be able to see on a screen how your brain or your heart, for example, respond to much slower, deeper breathing. Here, you can learn to boss a racing heart by kick-starting your parasympathetic nervous system – and then receive an immediate reading of how effectively you did it. With these treatments, instead of waiting for your body to take over of its own accord, you tell it to start doing some of the things it would do if the threat had actually gone away.

One of the easiest things you *can* control is your breathing, so you might start by inhaling deeply for five seconds, and then exhaling for a slightly longer amount of time. Slower breathing is one of the first bodily signs that a threat has receded, so your body can be persuaded that the tiger is no longer lurking. If you keep doing this for a few minutes while wired to a machine, you will soon be able to see the effects. For people who might be inclined to think that "obvious" relaxation techniques like this are for hippies, it's really helpful to have some hard-core technology in the room to persuade you that the effect on your body is real. Biofeedback treatments – and their technology-free counterparts, such as old-fashioned relaxation/meditation classes – can help people learn to control all sorts of bodily phenomena, including things as seemingly uncontrollable as skin temperature, sweat response, blood pressure and migraines.

What happens in vagus ...

The medical explanation for these responses is that deep breathing – and particularly a long, slow exhalation – stimulates the vagus nerve, which has parasympathetic control of your digestive system, your heart and your lungs. The vagus nerve is the communication system between body and mind, and the name "vagus" comes from the Latin term for "wandering": the vagus nerve is woven through the core of the body, like a wandering pathway, relaying messages from the brain to your organs in the neck, chest and abdomen.

For instance, it's responsible for telling your mind that you've eaten enough and should stop now. It's also the reason why you might suddenly lose your appetite on hearing upsetting news. By breathing deeply, you inject a load of oxygen into your bloodstream. Your vagus nerve interprets this as "my host's blood pressure is too high", so it gets on the case and causes your heart to slow down. It's in this sense that it's a "trick": instead of your mind telling your body to calm itself, it goes the other way around, and your body starts acting as though the mind has already done this, thereby fooling you into feeling calmer. Once you learn to set this chain of physical events in action, you may ultimately feel more in control of your anxiety, instead of the bodily symptoms of anxiety controlling you.

Other techniques for stimulating the vagus nerve include plunging into cold water, like the hairy Dutch "Iceman" Wim Hof – poster boy for becoming sage-like through freezing your butt off. Another is practising meditation, which not only encourages you to breathe more deeply and slowly, but invites you to physically be still, further persuading your body and mind into agreeing that nothing terrible is about to happen. Similarly, singing or chanting not only get your breathing going, but also activate your vagus nerve via your vocal cords and throat, to which it's closely connected. Belting out opera in the shower truly can help your body and mind to become better connected. Laughter, too, stimulates the nervous system into its soothing mode, which is why Laughter Yoga may be a fad that's not so laughable after all.

> I find that complete escapism in the form of watching old slapstick comedies or being silly with my daughter is the only thing that properly switches off my mind from mulling over climate stuff now. I always feel 100% better after having a true proper full-body laugh. Let the joy in! It's so easy to forget to do.
>
> Emi, teacher

It's a good idea to take deep breathing and meditation seriously. Deep breathing is about as ecologically harmless and freely available as it gets – you don't even need trainers for it.

The Pesky Human Mind

Still, if it were as easy as taking a few deep breaths in order to feel better about the climate emergency, then all of us would do it and that would be the end of that. The problem is, take a few deep breaths, feel better – and then what? The world remains unchanged. It's even possible that your mind will be so fixated on the horrors of climate breakdown that it won't give you that little bit of respite to breathe. While it might be scientifically proven that deep breathing can calm you down, it's also scientifically proven that terrible real-world situations make you anxious and stressed. You don't want to lock your body and mind into an endless one-person battle: *Calm down, I tell you … Why should I? The world is in danger … Calm down because I said so; I did a breathing technique … I don't care what you did, the world's still in danger …* and so on. The human mind is like the most renegade puppy: quite unruly and famously difficult to bring into line. You can only control your own thoughts up to a point. So while you might be doing all the "right" things to help yourself, you may still find yourself lying awake at three in the morning feeling helpless and infuriated about the failure of world leaders to protect the planet (or whatever your particular form of ecological self-torment leads you to obsess over).

Therapeutic Treatments for Eco-anxiety

This is where therapy can be helpful. The therapy most regularly recommended for people with climate anxiety is CBT. This is because CBT, and its mindfulness-based form, MBCT, is currently the most recommended form of therapy for everything. The reasons for this are that it's short-term, so is preferred by insurance providers and public healthcare systems, and because it is "evidence-based". CBT gives you quick, take-home, sensible techniques that you can use to correct your own "erroneous thinking". MBCT applies CBT principles to mindfulness techniques which are woven around cultivating an awareness of how bodily sensations link to emotional responses, so you can better understand and pre-empt your triggers for feeling anxious, and take steps to soothe them through cognitive practices rooted in CBT, e.g. "When I feel anxiety, I feel a rising tight sensation in my chest, I notice negative thoughts begin to swarm, multiply and catastrophize ..." The main flaw in CBT/MBCT as a "treatment" for climate anxiety is the theoretical plank it is structured upon: that "thoughts are just thoughts/thoughts are not facts". When you *are* dealing with undisputed evidence of habitat and species loss, raging wildfires, floods and global rising temperatures, this can be a problematic thesis upon which to base your reassurance.

CBT talking therapy can be incredibly helpful to people, offering them tools for dealing with the heat of the moment, which can unhook their minds from boiling over with whatever is bothering them, a kind of "pause" mechanism for an overly whirring brain. It can also give them techniques for in-between anxious times to help prevent the build-up that leads to a crisis. For some people that's enough, whereas for others it's annoying to be given little routines whose aim is somehow to placate you, when what you really want from therapy is to be able to explore difficult ideas, voice painful memories, and to find a space where the often severe difficulties of human existence are taken seriously. CBT is unlikely to be able to give you this. However, as therapists often like to repeat, research shows that *all* forms of therapy are equally effective. The real difference lies with the person who is the therapist: some people are just better therapists than others, but that's harder to explain within the "evidence-based therapy" model. If you think therapy might be able to help you, but you have no idea which form of therapy to try, it can often be a good idea to visit a couple of very different therapists and see who you like, or to try something short-term like CBT first and see whether it leaves you wanting to explore further, or whether it does what you need it to do.

Therapy or no therapy, when you're kept up at night worrying about the planet, it can be helpful to bear in mind two things:

1. Your own power is limited (even if you're a world leader). If you're doing what you can to stay informed, and to act responsibly in accordance with what you know, then it's unlikely there's some secret you've missed that would change everything if only you could think of it. Stressing yourself out does no good to anyone, especially not yourself, so you might as well give yourself a break.
2. If you're feeling *constantly* stressed about global heating,[1] it's highly likely that your complex human mind is doing all sorts of tricky stuff to *keep you in that suffering zone*. For whatever reason (it will be particular to you), you are almost certainly

> blending eco-anxiety up with all kinds of underlying psychic pain. Perhaps you feel guilty about bullying your siblings/ neglecting your mother/being generally selfish and rivalrous, so your caring attitude toward the climate is a way of attempting to redeem yourself (which also leaves you feeling that you're never quite redeeming yourself enough). Maybe you feel let down by your father, but can't take it out on him now he's old and frail – so instead you mentally castigate any number of slack, uncaring men in power positions, entirely justifying your unquenchable sense of childhood rage. Or maybe your mum was always totally absent and now you can't bear the idea of Mother Earth not being there for you either. I don't mean to parody these positions, nor to undermine people's seriousness with regard to their eco-concerns – for the record, I personally identify with all of these caricatured scenarios – but just to say that *psychic suffering is never simple*. Our minds are built to make connections, spot similarities, create analogies. We don't think in 2D. When we're trying to address human distress, our own included, we have to remember that we can't take everything at face value.

While not everyone will be inclined to do psychotherapy to work out what their unconscious minds might be contributing to their conscious worries, it's simply worth taking into consideration that, like all human beings, you are a knotty creature, somewhat cut off from nature and not always inclined to act in your own, or other people's, best interests.

Talk it Out

What is clear is that having *conversations*, whether as part of "talking therapy" or simply in someone's living-room over a cup of tea (the simple beauty of which we will never take for granted again having been through the social distancing lockdown of the coronavirus pandemic), or over Skype, is incredibly valuable. The main negative about experiencing eco-anxiety is the feeling of being alone in your fear; being the only person who cares. Finding people with whom you can share ideas and create a support network is invaluable. Being able to let out the words from your head, and give voice to your fears and connect to others who share your worries, can allow you to feel less vulnerable and more proactive, in whatever small way. And ultimately, beginning to articulate your feelings about the climate crisis will enable you to feel more confident in your voice, to be able to find it and begin conversations with people who may not (yet) share your anxieties – which is the way that we will be able to move toward changing people's attitudes and creating hopeful momentum and actual change.

You can read well-meaning literature put out by environmental charities telling you to breathe and walk in green space, and be nice to your neighbours – but still you may find it hard to put things in any kind of perspective; your thoughts never quite settle. This is very much to be expected. So while it might mean that yogic chanting doesn't have quite the magic effects on you that you might wish for, your "incurability" could also make you a tolerant, realistic and empathic person who's well placed to understand other confused, messed-up people, and to forgive yourself and them if anyone falls short of perfection. In other words, if you're doing whatever you can, but finding it's somewhat more confusing and complicated than that, *you're probably exactly the kind of anxious person the planet needs right now*. And if you're worried about the future, you're not alone.

ANXIETY: CHAPTER OVERVIEW

1. If you're anxious about the planet, you're right to be. We need more people like you! Stay anxious!
2. Moving your body helps to shift the mind's tension too.
3. Deep breathing can soothe you out of fight-or-flight mode – even machines say so. If you don't believe it, go for an encephalogram.
4. The other thing about deep breathing is that you can do it on public transport, but it's ill-advised to do it while driving a car. Win–win for the planet!
5. CBT, MBCT or psychotherapy can be helpful, as can simply being honest with yourself about the possible underlying causes of your anxiety.
6. Take action against the threat (see Chapter 8).
7. Talk about it: making connections, spreading awareness and creating ripples via conversation are some of the most powerful things we can do.
8. There's always medication, if that is the right route for you.

2
CRY, BABY: GRIEVE BUT NEVER GIVE UP

"It's possible to live in that deathly silent moment. It's possible to function. It's possible to do the most necessary thing, to ensure your own and your children's survival. It's possible to earn a living, buy groceries, cook dinner, do the laundry. It's possible to laugh. It's possible to have a nice time."

Naja Marie Aidt,
When Death Takes Something from You, Give it Back

In his broadcast on climate grief on his channel Philosophy Tube, the British YouTuber Oliver Thorn highlights the important distinction between grief and despair.[2] While the latter might make you give in altogether, the former is a dynamic process that somehow delivers you to a new place. For this reason, he suggests, it isn't necessary to lie to the public about the state of the climate. The presumption that we will fall into despair, rather than be mobilized by bad news, could be just another excuse to continue to try to hoodwink us: "Don't tell the poor dears – let's just keep it among us grown-ups. It would only upset them." Of course, climate deniers may *prefer* it if we disappeared under our duvets. A far greater risk would be that we might collectively revolt against their toxic activities: billions of individuals have immense power. The more you isolate yourself in your eco-anxiety, the more likely you'll fall into despair, and be less able to proactively help our planet out of her predicament.

This chapter will have a look at the terrible sadness and sense of loss many people already experience with regard to climate change. It will also suggest that there are ways of processing losses that leave you free to enjoy yourself, rather than hurling you into states of helplessness and melancholia. Perhaps it's even possible to value life *more* when you truly understand its fragility. So, rather than running away from the truth – or letting people in power conceal it from us – we can be realistic about the devastating ecological threats we face, and use this knowledge to deepen our love for, and connection to, our planet.

You Don't Know What You've Got Till It's Gone ... Or Do You Know Only Too Well?

We'll begin by exploring climate grief through the prism of Freud's theory of "pre-emptive mourning", developed in the wake of the First World War – a precursor of Lise Van Susteren's "pre-traumatic stress" (see the next chapter, page 45). In his short and beautiful essay "On Transience" (1915), Freud describes a countryside walk with a "taciturn friend" and a "young but already famous poet". Freud is perplexed to discover that his youthful companion is unable to enjoy the blooming foliage because it is "fated to extinction". (The people in question were likely to have been the psychoanalyst Lou Andreas-Salomé and the poet Rainer Maria Rilke.) They are, of course, not worried by climate change – we hadn't started wrecking the planet quite as much yet – just by the ordinary cycle of loss in the seasons, and by the more terrible notion of the eventual death of *everything* at some unknown point in the future. Freud tells us: "The proneness to decay of all that is beautiful and perfect can, as we know, give rise to two different impulses in the mind. The one leads to the aching despondency felt by the young poet, while the other leads to rebellion against the fact asserted." In case it sounds like Freud is anticipating Extinction Rebellion, the type of "rebellion" he's referring to is something more like "denial", as, in fact, everything *will* indeed end at some point. Still, he's alluding more to an old school, solar-system-inspired rock-crashing-into-sun ending, billions of years in the future, not to a major man-made fuck-up right now.

The inevitability of loss and death *needn't make us value things any less*: "A flower that blossoms for a single night does not seem to us on that account less lovely." In fact, limitations and inevitable endings might even increase the value and enjoyment of something. Freud seems to be arguing that *anything good or beautiful is worth valuing* – that's what life's all about. Being downcast about winter coming while summer is in full bloom makes no sense. We could also say, by extension, that neither does being endlessly miserable in the face of future climate change: mourning something before it's actually gone is perhaps the ultimate sign of admitting defeat.

Freud ponders: "Mourning over the loss of something that we have loved or admired seems so natural to the layman that he regards it as self-evident. But to psychologists mourning is a great riddle, one of those phenomena which cannot themselves be explained."

Why does grief have such a strange effect on people? For Freud, it all stems from a person's balance between self-love (or narcissism) and object love (being interested in other people). According to Freud, the love you feel for other people is borrowed from the more fundamental love you feel for yourself, as if each person contains a definite quantity of love that can be divided up and shared out, turned inward or outward according to whatever's going on in our lives at the time. For example, if you're getting loads of likes on Instagram, you'll have tons of narcissistic supplies coming in from the outside. This may mean you feel buoyed up and suddenly more able to be sociable and friendly to others in your "real life". If someone you love dies or leaves, the love that had previously been apportioned to them will be at a loose end. Of course, you don't stop loving someone because they vanish, but the love you felt for them will be changed: a running stream abruptly dammed mid-flow by their disappearance. Either you will have to redirect the love elsewhere, or it will need to flow back into yourself, ready to be re-released. Whatever happens, it's a time of transition which is likely to result in emotional suffering.

Grief is most problematic where there is a refusal to let go, and the love "flow" is eternally pushing against its dam of loss. Mourning can become despair, or melancholia if it never moves on but remains stuck in this initial state.

Halfway through his essay, Freud jumps from contemplating mysterious emotional processes to describing some of the horrors of the First World War, which "destroyed not only the beauty of the countrysides through which it passed [...] but it also shattered our pride in the achievements of our civilization [...] and our hopes of a final triumph over the differences between nations and races". Because of the level of destruction and the huge amount of loss experienced, many people all but lost faith in sophisticated cultural developments (such as moving toward acceptance across boundaries of race and creed) and clung instead to love of country, love of family, and whatever else felt easiest and closest. Freud believed this would be a *terrible* place to leave things. Just because the pacts between nations and cultures proved so destructible, it doesn't follow that we should now give up on them and become small-minded, jingoistic and selfish. He tells us: "I believe that those who ... seem ready to make a permanent renunciation because what was precious has proved not to be lasting, are simply in a state of mourning for what is lost." We could apply this concept to the current political atmosphere, and to climate deniers. Rather than deciding that they are simply idiots, we could cut them some slack by saying that they are somehow not ready to mourn their current way of life. They are clinging to the vision of a future that no longer exists, and are simply refusing to admit the truth of the situation to themselves. In essence: if ignorers of the climate crisis could let go of a future which revolves around capitalist notions of continual economic growth, consumerism, cheap flights, SUVs, mass-produced hamburgers, they would be able to invest in new ideas that may unlock a better future.

Jem Bendell is Professor of Sustainability Leadership at the University of Cumbria, and one of the outspoken heroes of the contemporary climate movement. Like Freud, Bendell has had to battle against his colleagues' disapproval, fighting to wake the world up to deeply unsettling ideas. His explosive essay "Deep Adaptation: A Map For Navigating Climate Tragedy" was written in 2018 and all but suppressed by less brave academics, before going viral as soon as Bendell made it available online. Much as Freud outlined the wastage and destruction witnessed in the First World War, Bendell confronts us with a visceral picture of the impending horror of societal collapse: "With the power down, soon you wouldn't have water coming out of your tap. You will depend on your neighbours for food and some warmth. You will become malnourished. [...] You will fear being violently killed before starving to death."[3] This fear of losing the future and the minutae of the lifestyle we take for granted – heated homes, running water, abundant food – is something that those of us experiencing climate grief know only too well. It is this *fear of loss* that needs to ripple through more people's consciousness in a powerful way, visualized graphically through essays such as Bendell's, in order for the climate crisis to ultimately galvanize the radical changes in human behaviour that it so needs.

Not Exactly a Walk in the Park

How can we allow the sadness in, without becoming swamped by the enormity of the losses we face? Of course there's no easy answer, but perhaps it's helpful to break it down into parts: loosely speaking, we need to stay engaged with the outside world while also taking our own internal worlds into account. We can't just do one or the other. If we throw ourselves into activism and ignore the fact that we are human beings with feelings and failings, we are liable either to run ourselves into the ground or to come up against resistances we don't understand: "Why do I always constantly crave meat/sleep all weekend when I could be out planting lavender/hate everyone in my local XR group?" And if we only focus on ourselves and our feelings, we will always be haunted by the idea that our concern for the planet is not altogether honest or altruistic. So here we will look at two very different approaches to climate grief – one that is political and one that is psychoanalytical – and begin to think about how the two might fit together. While we may never reach a seamless merging of the two strands (our outward ideals and our inner feelings are notoriously liable to clash from time to time), we can at least try to be tolerant of ourselves as we attempt to live our lives as un-idiotically as possible.

Oliver Thorn begins his Philosophy Tube broadcast on climate grief by speedily and persuasively connecting environmental issues with migration and labour rights. He does this using the example of fishing. Fish numbers are declining. Therefore it's harder to make money as a big commercial fishing company. One way to balance the books is by paying your workers less, and making them fish in more dangerous conditions (perhaps later at night, or further afield in harder-to-regulate, or even illegal, areas). In order to get people to do this without the risk of them reporting you, you will have to employ undocumented migrants. So the depletion of the oceans feeds labour rights abuses, which are made possible by stringent laws against migration. And migration is in turn fed by climate change and wars fought, funded and furnished by countries who close their borders against the very migrants they have helped to create. Wars and climatic disasters produce workers who can be used to further deplete the oceans, helping to continue the cycle of yet more wars and disasters. In other words, if you want to think about climate breakdown seriously, in ways that may actually bring about change, you can't just reuse a plastic bag and be done with it.

This includes internal change as well as external. As thc eco-psychoanalyst Joseph Dodds puts it: "Without the hope that meaningful, as opposed to manic, reparation is possible, we have only the choice between denial, madness and despair."[4] Mourning is a deep, complex process: offer yourself *false* hope about the climate, you will only get more depressed. No one wants to be the ecological equivalent of Charles Dickens's Miss Havisham (who was jilted at the altar, and who responds by wearing her wedding dress for the rest of her life), deluding themselves that if they stay on top of their recycling everything will magically fall back into place. If you want to feel better about the planet, and your place in the global system, a certain level of activism and change is arguably the only way forward. And not just environmental activism – you also need to think about border controls, labour laws, wealth distribution, arms trading, demilitarization, racism and all the other so-called Social Justice Warrior things that certain people might try to embarrass you out of. Or to put it another way: if you care about the world, you need to care about the *whole* world. In the end, you can't separate biodiversity from social justice; it's all connected, i.e. you can only burn huge swathes of rainforest if you completely disregard the rights of its indigenous human inhabitants – and all for the sake of beef-eating.[5]

So how's that supposed to make you feel any better? Perhaps the point is to take things out of your own hands a little, maybe to make it *feel a bit less terrible*. Ultimately, sitting around at home doing deep breathing exercises is only going to get you so far. The whole problem of climate-related suffering is exacerbated by the persistent background idea that it's down to each of us, on our own, to make our lives good. If we can't, it's because we're a bit crappy and should do some more work on ourselves. But what if we were living inside an extremely problematic socio-economic system that made it very difficult for us to "be happy"? A system that sold us the idea that "happiness" itself was a very desirable goal that could be achieved through, essentially, buying the right products? What if this system was perpetuated by greedy people who clung to resources for completely pathological reasons? And, furthermore, what if, in order for the rest of us to suck it up, we were coerced into accepting an ideology that made our unhappiness our own faults – the belief that if only we were clever enough to work out how to become one of the privileged few, all of our problems would be solved?

As a therapist, it wouldn't be appropriate for me to encourage people in my consulting room to think that these problems would all go away if only they could develop better personal habits and recycle more. This forcing of the solutions to the climate problem onto the individual is often echoed in well-meaning climate-anxiety literature which, as Jem Bendell puts it, "encourag[es] people to try harder to be nicer and better".[6] It's not a bad idea in itself, it's just unlikely to shake things up very much, and heaps on more anxiety about "not doing enough" to those of us who are already highly sensitive to what isn't being done.

"The challenge to me is on how to both take action and set realistic boundaries on what you can and can't achieve. Functional denial is necessary to some degree, otherwise we might be literally petrified, but finding the balance is tricky. Having had 3 percent of Tasmania on fire last year, then the accumulated extreme weather events of drought, mega-fires, storms and now floods all down the eastern seaboard of Australia, it is highly relevant to us. Alpine regions, rainforests and even mangrove swamps have burned.

As a GP, I find that those with existing mental health issues are more vulnerable to the added societal angst – the blow falls again on the most vulnerable in society. But it is also grandparents and great grandparents – even residents of nursing homes – as well as the young ones."

Dr Clare Smith, Australian GP

What Are Babies Like?

Way back in 1972, the American psychiatrist Harold Searles wrote, "[M]an is hampered in his meeting of this environmental crisis by a severe and pervasive apathy which is based largely upon feelings and attitudes of which he is unconscious."[7] What might these unconscious attitudes be? In order to answer this question, he delves back into the murky depths of our earliest experiences. All of us were babies once, and we are still shot through with whatever we experienced at the time, even though (or especially as) we don't remember it consciously. He explains that a breastfed baby is liable to experience her mother's breasts as both lovely and dreadful due to the fact that sometimes they're there when you want them, but sometimes they aren't, and sometimes they make you feel blissfully full and satisfied, while at other times they leave you feeling colicky. (You can loosely substitute bottle for breast, father for the mother, etc.) Because a baby is unable, as yet, to work out who's who and what's what, she might split her mother into "good" and "bad", experiencing her one way and then the other. She might also sometimes see herself as toxic or damaging due to the rage she feels toward her objects (the things that are other than herself). This is because she has no idea what the awful feelings she's experiencing *are*, although she may feel responsible for them. She might feel miserable as she grows up to realize that nothing involving other people ever seems to go perfectly – but that being alone isn't a great option either. Everybody, including oneself, is a perplexing mixture of good and bad, nice and nasty. And that's the best developmental outcome you can hope for.

Furthermore, even the sanest, realistically depressive adult will be haunted by the drama of their early interactions. So you may be liable to split others into good and bad, or to project your own feelings onto (or even into) other people, either misperceiving or controlling those around you according to a mode of relating developed before you had anything like a sensible cognitive map. You will almost certainly be riven by hatreds and rivalries, even toward those you love most, and you will mistrust those around you, presuming they're all secretly as bad as you are. In short, even the sanest humans have extremely problematic relationships with their most important objects, which includes everything from their parents, to their children, to their partners, to their possessions, to the world as a whole. Any of these entities might find itself positioned as "good", "bad" or problematically oscillating between the two.

Searles makes a number of observations about how these early relationships might inform our attitudes toward the planet. If the world is seen as maternal in its capacity to feed and comfort us, it can also seem to turn against us and cast us out, maybe when resources become scarce, or are plentiful but toxic. "Nature" and "technology" might be held up as either good or bad mothers: in the old days, "good" technologies such as architecture and medicine protected us from the evils of nature. But nowadays the situation is reversed: cruel technology is either poisoning or displacing us, and we need to re-find our lost "good" Mother Earth. Still, we're perhaps confused as to who to turn to for succour – which "mother" will take care of us better?

In terms of re-ignited rivalries, these might appear in the form of political struggles between classes; we might take out our vengeful feelings, enjoying the fact that the destruction of the planet will affect both rich and poor. (Although, of course, not equally, but maybe that was less obvious in 1972.) Then again, perhaps the nature of *all* our object relations has been so thoroughly debased by the atomized, selfish bent of modern life that we simply stop caring. As Searles puts it: "Is not the general apathy in the face of pollution a statement that there is something so unfulfilling about the quality of human life that we react, essentially, as though our lives are not worth fighting to save?"[8] Like sulky children, we don't mind ruining the family holiday because our parents are boring and distracted, and our brothers and sisters won't share their toys.

One of the many charming things about Searles is that he doesn't even remotely play the role of the haughty mental health expert, somehow above the fray. Alongside speaking about people's pathetic dependency on cars to give themselves a sense of power and independence (in order to kid themselves that they aren't in fact extremely vulnerable and highly dependent), he discusses his own temptation to jump out of a moving car on his way home from work: "I felt so shamefully and desperately unable 'simply' to face the living out of my life, the growing old and dying, the commonest, most everyday thing."[9] The pure facts of living in the world and not being able to control "living and ageing and dying", of having to work and to travel, of getting older, were an affront to his sense of omnipotence. One way of getting back on top of things, paradoxically, would be to take his own life; that'd show the world who's boss. Alternatively, you might demonstrate your freedom from the tragic difficulties of life and death by putting all your faith in science. But this too is an imperfect solution. Searles likens this to a person who becomes rich and famous, supposedly transcending all the banal hardship of the average schmuck, but who then discovers that they are disqualified from experiencing real human love. It transpires that it's actually horrible to actualize your omnipotent fantasies. Similarly, according to Searles, it would be unbearable to achieve full domination of our planet. He hypothesizes that we would rather see it destroyed, and ourselves with it, than to allow for the nightmare of post-human, post-planetary life (i.e. scooting around space freely like "gods or robots"). This "solution" is the most depressing of all.

Lastly and perhaps most importantly, according to Searles, older generations can take out their rivalrous feelings toward younger generations by making the planet a much more difficult place to live. This is one of those brilliantly counter-intuitive (at the same time as blindingly obvious) psychoanalytic observations that, to me, proves that analytic thinking is still vital to mainstream cultural life. While you hear all the time that "we must save the planet for the sake of our children", what you simultaneously see everywhere is people – who very often have several children – not doing that.

Basically, according to this theory, the earth is an object that we take our shit out on. (And the word "object" is very important to Searle's thinking, which is based in "object-relations theory" – a field of psychoanalysis that focuses on our relations to our important objects, a.k.a. the key people and actual objects in our lives.) To put it more nicely: "We project upon this ecologically deteriorating world the deepest intensities of all our potentially inner emotional conflicts." And that's why it's so difficult for us to get together and agree that it would be a good idea to start looking after the place.

Still, it's not as though we're completely helpless against our inner drives and conflicts. Society exists because people have found ways to band together. We can decide not to do bad things, or to do good ones. And if we know a little more about the unconscious forces underlying our conscious intentions, we may be slightly better guarded against acting unthinkingly, taking out our aggression unreasonably, or attributing ideas to other people according to our own warped perceptions. In other words, a good working knowledge of our own unconscious can help us to treat both ourselves and other people better.

A CASE OF CLIMATE GRIEF

A young man – we'll call him Liam – who comes to see me (and who is generous enough to have agreed to let me tell a fragment of his story) seems to me to have taken some quite amazing leaps in exploring the ideas underpinning his grief and anxiety about the state of the planet. Liam began weekly sessions a couple of years ago in order to deal with feelings of low confidence, and doubts about his relationship. Like anyone in long-term therapy, the subject matter of his sessions soon began to widen out into discussions of his parents, and even grandparents. It transpired that, although he had grown up with a sense of economic stability – even affluence – both of his parents had experienced poverty in their childhoods, his father in particular. In Liam's father's case, the material deprivations he suffered were matched by emotional ones – he described his childhood as having been loving but chaotic – you never quite knew where your next love might be coming from.

As Liam grew up and left the family home it seemed that his father was becoming more and more obsessive about the prevention of waste – to the point where it was extremely uncomfortable to go back to visit. The father's anxiety about water, draughts, vegetable peelings, etc. was so pronounced that a meal couldn't pass without a stressful incident; someone would do something 'irresponsible' (like leave the back door open for too long) and the atmosphere would turn sour. It was never quite clear whether the father's concerns were environmental, or economic, or whether he just had a terror of anything vaguely uncontrolled. In fact, he seemed to bundle all three options together into a potent kernel of outwardly projected angst.

Liam was in a long-term relationship with a woman he'd met at university, but could never quite commit to her. Part of the reason for this was that he couldn't imagine a future with her, and especially not a future with children in it. She seemed to want them, although in no particular hurry. However, he was sure his father would disapprove if she got pregnant. While Liam's father had never explicitly discouraged him, he would sometimes speak scathingly about overpopulation – linking this back to the horrors of being brought up in a family with eight children. It seemed that the whole world was just an expanded version of Liam's father's family – too many people and not quite enough of anything to go round – and that it would be irresponsible to keep replicating that. Therefore Liam felt unable to love his girlfriend properly, knowing that he would have to leave her if she seriously wanted kids. And, more than that, it opened up questions about whether his father had wanted to have *him*, whether he was just a burden – an idea that seemed to link itself to his friendships, and indeed to all social situations, even to his very existence. In other words, Liam had a tendency to see himself as a complete *waste of space*. This made it difficult to find meaningful work, have fun, or even to give himself over to altruism. Everything seemed fake, disappointing and pointless.

Liam would lie awake worrying about fires, floods, bugs (I actually think we might need a special term for climate-related insomnia, as so many people seem to speak about it) and also about the fact that his forays into activism were so half-hearted, at least according to his own stringent standards. The fact that he felt so awful so much of the time seemed to send him into a self-fulfilling misery spiral; what's the point in a person who doesn't even like life that much? How dare he go around eating, breathing, and leaving back doors open, and not even feeling particularly grateful for the resources he was so joylessly using up?

Over time it became possible for Liam to begin to untangle some of the threads that were keeping him unhappily entwined. While the climate emergency was real, it wasn't fully analogous with his father's experiences and attitudes. Sure, wasting resources is bad, but if you bind it in with the traumatic echoes of a chaotic, materially unstable childhood – and its unsettling, life-long after-effects – you get something properly dreadful.

By linking his own very real concerns about climate change to his father's unresolved childhood woes, Liam could start to separate things out a little, gradually finding himself freer to make choices about how he wanted to live, and perhaps to feel a little less responsible for his father's happiness or unhappiness. Of course he may still choose not to have children, to remain vegan, to own only two pairs of shoes, and so on, but without the burdensome unconscious overlay of his father's unresolved emotional difficulties.

Let's Not Object to Discussing Politics

Another popular use of object-relations theory is in the discussion of people's personal politics. For instance, how might your early relationships inform your attitudes toward sharing, your sense of trust in the world to sustain you, or of what you think it owes you? Do you believe you're all alone and have to fight for every scrap of sustenance? Or do you sit back in the conviction that good things will surely come your way? Perhaps your tendencies toward capitalism or socialism aren't purely the product of conscious moral choices, nor simply a reflection of your family's beliefs, but are also informed by more deeply held, unconscious ideas about the bonds between yourself and others, what's inside you and what's outside you? Or, to be really reductive, your relation to the breast. In short, what are you trying to do with your objects? What do you think they're trying to do with you? How do you get satisfaction? Or fail to? What makes life good?

All of which brings us back to Freud, Rilke and the question of pre-emptive mourning. Freud is arguing for a mode of being where we are able to bear our own trajectories toward death, and the death of everything around us, without throwing the baby out with the bathwater. It's in that bit there, where everything hangs in the balance, that we can experience joy, immerse ourselves in love and beauty. With a nuanced relationship to the things we love, and the things we risk losing, we can coexist, share, receive gifts, mourn and enjoy the present in all its transient glory. A high level of cooperation and mutual respect is what makes human cultures so awe-inspiring. In this delicate space, you have to be prepared to accept losses at the same time as being up for doing the work of recovery. *You must never give up before the bad thing happens.* And that way, you lay the ground for yourself and others to experience the unlikely wonders of existence.

GRIEF: CHAPTER OVERVIEW

1. Being sad is not the same as being hopeless – it can be a step on the way to something much better. Allow yourself to feel sadness in order to begin accepting it and so that you can start to see what may be there when you come out the other side.
2. Try not to pre-empt your losses in advance: truly savour what you have right now, in front of you.
3. If your worries about the climate seem to connect with worries about all sorts of other things, you're probably on the right track. It's not crazy to make the link between wars and tuna sandwiches via displaced migrant workers.
4. All of us were babies once, and it takes quite a bit of getting over. No one ever fully grows up. Don't worry if your moods can sometimes seem a little immature, that's just life.
5. If you can, ask your mum about your earliest experiences. It might help you to think about why the climate situation affects you in the way it does.
6. Try not to think of the word "loser" as an insult. Being good at losing things is a valuable skill. In fact, the urge to be constantly seen as a winner is causing major personal and planetary problems.
7. Avoid walks with grumpy poets.
8. Never give up on the world, or people, ever.

3
PRE-TRAUMATIC STRESS: CHEER UP, IT MIGHT NEVER HAPPEN (BUT THEN AGAIN IT ALREADY *IS* ...)

"Imagine a future. Be in it."

Björk, singer-songwriter

It might feel like over-egging the pudding to bring trauma into the discussion: upping the ante from anxiety to trauma may seem like a big step for the "everyday" experience of climate-related worry. But it's worthy of discussion, becoming a more present term within the emotional landscape of the climate crisis, as we wade deeper into our changing existence on the planet. It may be said that the wildfires of early 2020 in Australia invited a collective shock and trauma globally as we witnessed the effects of climate heating unfolding terrifyingly on a constant loop in our news feeds, in scenes previously observed only in a disaster movie. This chapter explores the definitions of trauma and how they might relate to "regular" eco-anxiety in this changing world that we're living in.

Up until the last decade, it was pretty much agreed that the word "trauma" necessarily related to events in the past. Trauma, in its psychological sense, loosely meant something that had taken place that was somehow too much to be seamlessly integrated into a person's experience. Either it was too sudden, too intense, too horrifying or too unexpected for it to be included in the normal run of events. Because of this, it left scars, traces; it proved difficult – sometimes impossible – to recover from. You simply couldn't forget and move on.

"Trauma" comes from the Ancient Greek word τραῦμα, meaning a wound or hurt, or even a defeat. By the seventeenth century, it was used in medical Latin to mean a physical wound. But at the end of the nineteenth century, it was taken up by researchers into psychology and applied to the mind; as well as its original meaning, "trauma" could now also denote some kind of psychic wound. Famously (or infamously), Freud developed the idea of childhood trauma to explain people's mysterious, non-physical symptoms. Since then the term has been associated with all kinds of human experiences, big and small – from wars to natural disasters to bereavements to romantic disappointments – with survivors of these events regularly being diagnosed with post-traumatic stress disorder (PTSD). Now this term is often used to describe the suffering of climate scientists who are seeing the effects of climate change first-hand as they emerge, and, more importantly, predicting how these might worsen in the near future.

What is PTSD?

Symptoms of PTSD vary hugely from one person to another. Perhaps the most recognizable forms involve flashbacks, intrusive thoughts and nightmares; memories of the distressing event repeatedly reappear of their own free will, wreaking havoc in the person's life. While this is obviously incredibly upsetting, the one upside may be that the disaster at least appears as *itself*.

Things become more complicated when the traumatic event is brushed away or pushed out of the mind, as if it were possible to live as though it never happened. While it might be tempting to think it would be better not to let this thing define you, this can lead to a situation where it becomes impossible to speak or think about the event directly, but the problem has far from vanished. Instead, it can begin to appear in myriad other forms, from angry outbursts, to addictive behaviours, headaches, insomnia, inability to hold down a job, free-floating anxiety, depression, phobia, fainting, OCD, panic attacks, physical tension and pain, and anything else your poor organism can come up with to force you to recognize that something doesn't quite feel right. Bessel van der Kolk's powerful book *The Body Keeps The Score* explores this in detail.

Pre-traumatic Stress

Since around 2015, a new possibility has begun to inch its way into public consciousness: the idea that one could also be traumatized by events *in the future*. The term "pre-traumatic stress" was coined to explain climate scientists' responses to the changes they were seeing in the environment. Not only were many of the most gloomy predictions of the last three decades – sea-level rises, land erosion, animal extinctions, wildfires – coming true, sometimes they were proving to be underestimates. This often brought about the further realization that things were therefore sure to become considerably worse in the not-so-distant future. The effect on the people witnessing and measuring these changes was unsettling. The scientists were beginning to report feelings of anger, distress, helplessness and depression. The term "pre-traumatic stress disorder" was popularized by the forensic psychiatrist Lise Van Susteren, who has since said she'd rather not use the word "disorder" because it's a *reasonable* response to a dire situation; the word "condition" would seem more appropriate.[10] One of the most plain-spoken scientists to raise the alarm was the climatologist Jason Box, who in 2014 famously tweeted, "If even a small fraction of Arctic sea floor carbon is released to the atmosphere, we're f'd." With this one small sentence he unleashed a torrent of reactions. His tweet went viral, and the world's media were suddenly on his tail. On the one hand, he was actually saying what many people had already concluded: things were probably far worse than we'd thought. The difference was that he was a scientist and he was saying it in the most robust terms.

The problem with much scientific writing on the climate was that it tried very hard to be unemotional and objective, but this meant that the writers often found themselves putting forward the most apocalyptic ideas in the coolest terms, afraid that any hint of panic on their part would undermine their credibility. The last thing the scientists wanted was to be accused of being hysterical or alarmist. On the one hand, it might make people less likely to trust them, not to mention risking putting them in the firing line of corporations who saw the honest reporting of the science as an attack on their ways of conducting business. On the other, it might make people give in altogether: "Oh well, if the scientists say we're fucked, we might as well have a party on the way out." All of which apparently left many of the scientists feeling understandably freaked out; they were the very people in a position to do or say something that might actually make a difference, but they were being thwarted at the last step, unable to articulate clearly the effects they were seeing unfold before their eyes. Another perfect example of the madness built into the global reaction to climate change. You couldn't simply say what you were seeing because there was apparently no longer any such thing as a simple fact. The facts were already political. If you dared to suggest that people might need to drive less, fly less, frack less and so on, you risked being called a controlling, envious leftie with no place in the hyper-modern, super-free, fantastically fun world. From here, climate change itself could be magicked into a big political conspiracy whose aim was to rob the rich. No wonder some of the scientists started to fear that they might be losing their marbles.

> "I flit between trying to convince myself things are 'going to be ok', to feeling utterly and completely terrified. My 4-year-old is at the age where he talks about what he'll be when he's a grown up or (heartbreakingly) that one day he'll be a Daddy when he has children … and my brain immediately wonders if that future is even an option for him any more. It is an intense vision of future loss, which sometimes feels unbearable."
>
> Becca, designer

Then there was the classic problem faced by scientists around funding: the source of your money is liable to affect the apparent results of your research. Of course, this isn't to say that the scientists were being paid to make things up, just that all data is open to a variety of interpretations, and if you're being asked to extrapolate predictions from the numbers, you can choose whether to be cautiously optimistic, to go right down the middle, or to be frighteningly pessimistic. While you could argue that taking a middle path is most likely to result in your being "least wrong", it may also turn out to be the case that the most extreme predictions, at either end, turn out to be right. They may even turn out to be underestimates, as we have seen from people's failure to account for the fact that icebergs also melt from underneath. (Original calculations regarding icebergs presumed they were melted from above, by the sun, not so much from below, by warming sea water. It wasn't until the last decade that people began to see that things were far worse than predicted.)[11]

The problem for everyone else was that the information has, until the recently labelled "climate emergency", filtered through in fractured, compromised and contradictory ways. We non-scientists can feel like the children of parents who are trying to get divorced without the kids noticing. We could tell that something was up, but the full extent of it wasn't quite clear. The older amongst us had been hearing about all this stuff since the 1980s, when hairspray and refrigerators suddenly became the main bogeymen due to chlorofluorocarbons, or CFCs, which were released into the atmosphere, causing a hole in the ozone layer. But since then, they'd fixed that, sort of, and it was easy to lose track of how things were shaping up. One day you'd be told that the ozone layer was on track to complete recovery, but then the next day you'd hear that huge cracks had opened up in an Arctic ice sheet. It could be hard to tell whether things were largely improving or declining. Climate change deniers could occasionally sound very convincing, presenting data that made you doubt your own beliefs – what if yours fears were rooted not in facts but in your flaky, neurotic temperament? There's nothing like another person's absolute certainty to make you question your own position. If you're each equally sure of mutually exclusive ideas, you can't both be right. And if the other person appears more vehement, better armed with figures, more loquacious, then you may very well be tempted to cave in.

Climate change denial is coming to seem like a spectacular form of gaslighting, i.e. making people question their own sanity by lying to them (a particularly appropriate analogy in that it features the misuse of fossil fuel). Even the deniers seem to know that their overall arguments are unsustainable (literally) and that they are only putting them forward as a tactical manoeuvre to support their own – or their financiers' – immediate monetary interests. All of which is to say that smoothly processing information about climate change is practically impossible, so the likelihood of being left feeling discombobulated, uncertain and freaked out is high.

Cli-fi

While the science behind human-led climate change acceleration becomes ever more unarguable, it's uncertain what it will mean for us as people. Will we be forced to band together to defeat this terrifying enemy? Will it make us *less* caring, more cut-throat in our eagerness to snatch at dwindling resources (much like the empty supermarket shelves around the COVID-19 outbreak seems to suggest)? Or will humanity divide along these two lines, resulting in a *Game-of-Thrones*-style battle between good and evil? Oh, but of course, that's already happening ...

While the scientists may have been reacting to the environmental losses they were seeing unfold, they also spoke about the fear of the impact on human society as a whole. The level of catastrophe they foresaw was so high that it couldn't fail to impact on people's thoughts, feelings and behaviours. Since 2013, when the American radio station NPR coined the term, we've had the word "cli-fi" to designate and categorize films and novels dealing with eco-disasters, which people had already been making for many years by then. Films such as *Blade Runner*, *A.I.* or *The Day After Tomorrow*, or books such as Ian McEwan's *Solar* or Margaret Atwood's Maddaddam trilogy, could all be said to fall into this category. Another less obvious example might be *Terminator 2*, which depicts a world where machines have taken over and are planning a nuclear holocaust. The heroine, Sarah Connor, is the ultimate sufferer of pre-traumatic stress. Much like a contemporary climate scientist, she can see where things are headed but is having great difficulty persuading other people to take her seriously. In fact, the more angry and insistent she becomes, the more she is punished for it, creating a feedback loop whereby the less she is believed, the more she needs to harangue people – but then this vehemence is taken as further proof of her madness. Her unshakeable knowledge that she is right actually works against her, resulting in her being locked up in a psychiatric hospital. Not everyone who has an unswerving conviction about future events is correct – catastrophizing around apocalyptic scenarios has always been one of the typical symptoms of psychosis. Anyone who predicts the end of the world as we know it leaves themselves open to accusations of madness. Of course this is a false deduction; as Joseph Heller pointed out in his novel *Catch 22*, "Just because you're paranoid doesn't mean they aren't after you."

Greenland

In August 2019, the same month Donald Trump offered to buy Greenland, the Greenland Perspectives Survey was published. A collaboration between the universities of Copenhagen and Illisimatusarfik, it was a huge, multi-disciplinary project that took Greenland to be a kind of test case for the effects of climate change. Because of the country's location and particular climactic conditions (basically, it has loads of ice – for now), it has found itself at the frontline of global heating, registering changes sooner and more visibly than countries with more temperate climates. Perhaps unsurprisingly, the survey found that more than 90 percent of Greenlanders agree that climate change is observably happening. They can see the effects of it all around them: the storms are weird and different; they last longer and are more intense; their landscape's changing, and therefore animal behaviours are affected. For instance, polar bears are more likely to invade villages and attack people as they have to look for food in different places. Because of the altered environmental conditions, economic conditions are also affected, so survival is much more difficult; people used to be able to find food but now they have to buy it, which they might not have the money to do because it's not a system they're used to or equipped for. They may not be able to see friends and relatives because the ice, which they used to be able to travel across in winter, is thinner, so it's too risky.

Apparently many Greenlanders are being forced to give up their sled dogs because they can't afford to feed them, and also because they can no longer use them for travel due to the thinning ice. People are more isolated and have fewer resources. Greenlanders aren't *imagining* a scary future, they're already in it. Unsurprisingly, perhaps, it's making them anxious and depressed; alcoholism is on the increase, and their suicide rate is rising. According to Kelton Minor, the lead author of the survey, "The Arctic is a bellwether for the unequal impact of global warming on social and economic systems. As countries struggle to limit future risks and overall warming to 1.5°C, many Arctic and Greenlandic residents are already living in regional climates that have changed by more than this, in less than a lifetime [...] Therein lies the paradox: while satellites and sensors monitor the surface of Greenland's ice sheet, chase icebergs and scan sea ice daily, relatively little is known about what the residents of Greenland think about their changing surroundings."[12] As it turns out, it's already a mental health catastrophe.

Around the world, people are finding useful new words and ways to describe the mental suffering linked to environmental change. There's climate-/eco-anxiety, climate grief and climate-related PTSD. Alternatively, the Australian philosopher Glenn Albrecht coined the term "solastalgia" to describe the anguish caused by environmental alterations due to droughts and destructive mining in New South Wales. Taking the Latin word for comfort and consolation in times of distress (*sōlācium*) and the Greek root designating pain (*-algia*), he invented a new word that sums up the awful effects of finding discomfort where you used to look for comfort.

"The hardest part of eco-anxiety for me is that the habitual witnessing of nature, which once was a completely magical experience, now has a bitter taste. Spotting the first snow drops of the year, watching the autumn leaves turn, collecting conkers; all these things have lost their innocence, since they are tainted with worries. Are they too early? Are they too late? That perfect connectedness is something I will never get back."

Ella, editor

In parts of Greenland, they have the word *uggianaqtuq* to describe the horrible feeling caused by a friend behaving strangely, or even a sense of homesickness experienced when you're actually at home. Recently, this word has been co-opted to describe volatile weather conditions and the sense of your surroundings becoming unreliable. It seems particularly apposite in relation to the idea of pre-traumatic stress. In cases of psychological trauma, one of the key precipitating factors is a break in the person's sense of trust. This could cover anything from a person who has been in an abusive relationship and who then finds they can't recover from it enough to trust another partner, to someone who has been in a car accident and can no longer believe that driving is safe. For people in Greenland experiencing huge socio-economic shifts due to environmental alterations, there may very well be a traumatic sense of loss of trust. The world around them might still appear similar enough for them to expect things to be roughly the same, but then, in an uncanny twist, it turns out that everything is subtly altered, just enough to make their old lives impossible.

This sort of breakdown of trust may also be the case for the climate researchers, both in terms of unsettling changes to the material world, but also a loss of trust in people in power, who are supposed to act responsibly toward the societies they regulate and inhabit. Why do they so often seem to ignore what the scientists are urgently trying to tell them? Shockingly, there are already luxury "climate-proof" apartments, with high-tech fire and flood-proofing, to protect the super-rich in big cities from catastrophe.[13] Still, it's not like you can survive in your sealed-off apartment forever while the rest of the world gets swept away. With unjust safety-cons like this around, it's perhaps not so odd that less responsible money-heads continue to kid themselves that they can harm the environment without it having dire consequences for them personally.

There are enough reasons to accept the notion of pre-traumatic stress, even if the idea has something slightly counter-intuitive about it. If psychological trauma is characterized by the inability to incorporate thoughts, memories, ideas and experiences smoothly into a functional and soothing sense of reality, then it's not a total lack of sophistry to say that the possibility of an extremely frightening near-term future might have traumatizing effects. How are we supposed to synthesize our ideas about the future into our daily lives when the information we receive about it is a weird mixture of scary and contradictory? (And now, in the midst of the COVID-19 pandemic, who knows what the state of "the future" will be by the time anyone comes to read these words. Perhaps those Greenlanders' problems will sound mild by comparison. I hope not.)

Trauma-proofing

Throughout the twentieth century, as Freudian thinking became popular and widely disseminated, psychologists such as Jean Piaget, John Bowlby and Michael Rutter came up with further theories around childhood development, which led in turn to the wishful idea that you could somehow trauma-proof your child. You would do this by having a really good handle on developmental theory, and by making sure you did everything by the book. You'd wean and potty train your kids at the optimum moments. You'd be there to guide them, but not too much. You'd give them all the information they needed to get on in the world (insofar as you knew it yourself). And you'd communicate with them with a perfect balance of realism and protectiveness. As it turned out, this didn't save your children from having nightmares, tantrums, trouble with friendships and all the other things that make childhood challenging – let alone failing to cause them to grow up into perfectly well-balanced adults. The tragedy of all this new psychological research and speculation was that it *couldn't protect your kids from the pain of the human condition*. However much you knew about the things that fucked you up, you couldn't save your loved ones from a similar fate.

Likewise, it would be hard to come up with a programme that could trauma-proof either children or adults against the threat of ecological breakdown and its accompanying societal cataclysms. It can sometimes almost seem that "sensible" advice can make things seem even worse. Like the state-sponsored, four-minute-nuclear-warning pamphlets in the 1980s, what could be more terrifying than a big authority telling you to stay calm when the bomb drops? However, just as one might not want to argue for a return to pre-Freudian ideas about child-rearing – i.e. lie to children to protect their innocence, and punish them if they step marginally out of line – it seems a no-brainer to say that it's preferable to be straight with the public about the state of climate change, and not to punish us if we then get pretty angry with governments and corporations.

Get Ready for the Future

Not everyone who has a horrifying experience will be lastingly traumatized by it. This fact has long interested psychologists, who have tried to uncover what makes the difference. Some of the answers are blindingly obvious, while others maybe less so. In 1983, the National Vietnam Veterans Readjustment Study (NVVRS) looked at the lasting effects on soldiers. Fourteen years after the war, they found that 11 percent were still suffering from PTSD. Looking more closely at the differences between sufferers and non-sufferers, one of their initial observations was that the presence of PTSD tended to correlate with the severity of their experience; the worse the event, the worse the aftermath. So far, so obvious. But when they delved a little further they began to uncover other variables. People who had experienced abuse during childhood were likely to be worse affected. This was perhaps due to the greater difficulty of feeling safe in any surroundings. Also, those with pre-existing mental health issues tended to be less resilient. Then there was the fact that younger soldiers were more likely to be traumatized, perhaps due to their increased tendency to extrapolate negatively from bad things that happened to them; older people might have had more experience of things panning out OK, so had a tendency to see disaster as the exception rather than the rule. Lastly, one of the key variables lay in the social networks surrounding the individual. People who felt supported by family, friends and communities were far more likely to fare well than people who were abandoned to their own devices. All of these variables will apply equally to people suffering from climate anxiety. If you have a pre-existing mental health issue, don't trust your surroundings, and have a sparse or fragmented social network, it may be far harder to process difficult thoughts and feelings around climate change.

Since the First World War, many therapies – beginning with psychoanalysis and ending with CBT and Eye Movement Desensitization and Reprocessing (EMDR) – have been developed to help people process traumatic experiences. Talking therapies largely involve giving people the time and space to articulate something about the things that have happened to them, at the same time as acknowledging that there will always be something "too much" – elements that elude description. CBT encourages people to retrain their thought responses to various situations and stimuli, and EMDR is basically a form of hypnosis with a sciencey-sounding name, to complete the *physical* response to trauma within the body. All of them can be helpful. But where might all this leave us with regard to the climate? It seems that many of the conclusions of the Vietnam War research are mappable onto the climate situation.

Certainly Greenlanders, Australian farmers (who are increasingly unable to grow their usual crops due to droughts and fires) and environmental scientists seem to be particularly deeply affected, as their experiences of it are more intense. The Vietnam survey also found that more traumatized veterans weren't just suffering from flashbacks and nightmares, but were more likely to succumb to addictions and to experience marriage difficulties and trouble in their relationships with their children, not to mention simply reporting greater levels of unhappiness than the non-traumatized. These descriptions of everyday unhappiness also surfaced in the Greenland Perspectives Survey, which uncovered increasing levels of depression, suicide, violence and alcoholism, presumably not helped by the fact that maintaining one's social and family networks is far more difficult when one can no longer move around so easily. As with the war veterans, younger people are particularly affected – although in the case of pre-traumatic stress, it isn't simply that your lack of experience leads you to be more likely to catastrophize, but also that you will bear the brunt of this particular catastrophe while the less-panicked older people are resting in peace.

Can We Trauma-proof Ourselves?

Perhaps one of the possible answers is that, while trauma-proofing may be impossible, you can help yourself a bit by doing all the things that help stave off mental anguish, such as eating as healthily and enjoyably as you can, allowing yourself proper rest (our immune system doesn't work effectively while in constant fight or flight), seeing people you like and can share your worries honestly with – but more importantly that you can also belly-laugh and totally forget your worries with – and so forth.

But it's vital to accept that there isn't a "correct" level of activity with regard to the climate – that each person has to gauge what they can bear. To under-prepare or over-prepare can seem like madness, from complete denial to learning martial arts in order to be able to fight when things get totally *Mad Max*. As the anxious parents of small children so often discover, the more you try to get everything just right, the more you risk making yourself and everyone around you miserable. You need to be allowed to think and speak about the worries you have, to act on them to an extent that you don't utterly destroy yourself and your relationships in the process, and to keep yourself informed without allowing yourself to be swamped by the horror of it all.

Quiet Power

There also needs to be a place for those of us who may not have a warm, wonderful extended family living down the road, or who may not always enjoy spending time in groups. There is some irony in the idea of curing climate-related suffering with sociability, given that it's our very passion for getting together in bigger and bigger conglomerates – whose deep-rooted aim is to keep us safer so we can live longer, breed more successfully and so on – that's got us into this mess in the first place. While there's plenty of scientific and statistical evidence to support the idea that collectivization is good, and friendliness staves off diseases, surely there has to be a place for quiet introverted people too? In his novel *Shy Radicals: The Antisystemic Politics of the Militant Introvert*, Hamja Ahsan makes fun of the "extrovert supremacy" who lord it over shy people, and who always presume that sociability is superior.

There is a risk that much of the mental health advice floating around at the moment takes a slightly extrovert-centric position, assuming that joining – or even organizing – groups, and generally being part of some shiny, happy eco-crowd is the only way forward. While this might be ideal for some of us, others of us might see the radical potential in quietness and solitude. For a start, it makes you far less likely to invest in fast-fashion, wasteful party food, and driving or flying around making sure your friends still love you.

Last of all, perhaps it's helpful to follow the wise advice of Björk and to counterbalance your gloomier predictions with optimistic ones. There's no point being fatalistic – *there's still time to act*. By simply allowing yourself to believe in a world where people are prepared to care for the planet they live on, perhaps you are already on the path to making it actually happen.

TRAUMA: CHAPTER OVERVIEW

1. Don't let anyone dismiss your worry about things that haven't happened yet; channel *The Terminator*'s Sarah Connor (but not too much).
2. People will be affected differently by the same events. The good news is that being traumatized isn't a sign that you care more; it just means you're more susceptible to trauma. So if you're *not* traumatized, don't worry – that doesn't make you Donald Trump.
3. The bad news is that susceptibility to trauma isn't necessarily something you can decide for yourself – unfortunately, you can't choose your own childhood.
4. You're not a robot. If you feel crap sometimes – even a lot of the time – it's part of the human condition, so don't beat yourself up for feeling bad.

5. Remember that uncertainty about the future is nothing new. All lives always have been, and always will be, unpredictable. That's just the nature of existence.
6. Don't try to trauma-proof your children; just love them in the best way you can, this is the greatest form of "keeping them safe".
7. Be as sociable or reclusive as your temperament dictates: generic advice is all very well, but you know yourself best.
8. Having said that, try to get out of your comfort zone sometimes. Extroverts: explore the ecological potential of a quiet life. Introverts: see if you secretly like hanging out with people sometimes.

4

DENIAL: STRICTLY FOR THE BIRDS

"Modern man is drinking and drugging himself out of awareness, or he spends his time shopping, which is the same thing."

Ernest Becker, *The Denial of Death*

In this chapter, we will consider the possibility that the denial of death has an important part to play in people's mistreatment of the planet. Denial is a perplexing subject – it can be baffling why some people seem unable to accept the idea that we currently have a problem on our hands with regard to the climate. Earlier, we considered the suggestions that they may not be able to bear the grief, that they may be so upset about the state of their own lives that they take it out on the world around them, or even that they don't have any love to spare for the planet because they feel so unloved themselves.

Rather than demonizing climate deniers, in this chapter we will try to understand them better – especially so that we can get to work on them, bringing them round to more planet-friendly ways of thinking. In order to do this, we will take a look at the work of an ornithologist, a cultural anthropologist and two different groups of experimental psychologists – one lot who work with fear, and another lot who deal with persuasion. If we are to have any hope of tackling other people's resistance to taking action on the climate, we have to become artful and open to inventive and unlikely tactics. Bashing people over the head with our opinions is liable to make them retreat further into their own preferred beliefs.

Everyone knows the famous vegan joke: *How do you spot a vegan? Don't worry, they'll tell you.* No one wants to be *that* vegan, so here are a few thoughts on eco-wiliness to help you win over the most impervious deniers in your lives. But first we need to talk about what's making them like that …

The Temptation to Deny

In January 2010, the *Cornell Chronicle* published a short article by the ornithologist and avian conservationist Janis L. Dickinson, a professor in the Natural Resources department at the university. In the first paragraph, she makes two predictions regarding climate change: firstly, that people will become far more entrenched in their worldviews – either more religious and fundamentalist, more frantically materialistic or more stridently eco-conscious. And secondly, that, "A fearful public will become more easily manipulated and deluded into a false sense of security or salvation."[14] Some might say she perfectly summed up the second half of the 2010s, with powerful countries split by seemingly intractable ideological differences, and a sudden influx of conservative, paternalistic leaders. How does an ornithologist come to make such striking predictions about human beings?

Dickinson's forecasts have their roots in the work of Ernest Becker, an American cultural anthropologist whose book *The Denial of Death* won the Pulitzer Prize in 1974. In it, he argues that human beings' fear of personal extinction is our primary motivating force – everything we do is at the service of avoiding death (even things like smoking, perversely, which I'll explain shortly). According to Becker, as our prehistoric brains bulked up and became more sophisticated, we developed language, and thereby the capacity to imagine and represent various futures. In other words, we became self-aware – which also means *death*-aware. As Eleanor says in the US comedy series *The Good Place*, "You're learning what it's like to be human. All humans are aware of death. So … we're all a little bit sad." And this applies to all forms of death – including planetary death, not just our own.

As a result, we had to invent the means to deal with this horrifying knowledge. In the past, this meant developing myths and worldviews that supported ideas like life after death, gods who protect humans, history, heritage, heroism and all the other things that give people a comforting sense of legacy and consistency. Rather than seeing yourself as a random, fleshy phenomenon, the trick was to tell yourself that life had meaning – and that your existence was somehow included in that. Becker's name for the outcome of this attempt at damage-limitation was "self-esteem". Rather than seeing self-esteem as the inalienable right of anyone who reads cheesy magazines, for Becker it was more an understandable attempt at self-delusion, or suspension of disbelief. Self-esteem was what promised to buffer you against an awareness of your own mortality; its aim was to keep you in a comforting state of ignorance – your life *means* something, it's important, no doubt the universe is unfolding as it should, etc. This strategy has surely been extremely helpful to homo sapiens over the millennia – allowing us to keep going with the analytical thinking, trying to understand the world around us without being forced to exist in a permanent state of existential dread.

But, according to Becker, it comes with an unfortunate backlash. Because humans have a great deal of previous when it comes to holding ourselves and our societies together with cosy illusions; we often prefer to turn to these in moments of serious threat, rather than coming up with more hands-on, real-world solutions. In other words, we rely heavily on self-esteem-based immortality systems. In practice, this might mean buying lipstick, a coat or a car in a moment when the future looks sketchy – perhaps when your neighbouring village is flooded; or when the world is swept by panic about a new virus; or voting for someone who promises to return us to some kind of Golden Age; or putting all of our faith in a spiritual higher power. This is all armour for self-preservation. Take this real life example, from a woman I work with who has a morbid fear of tunnels. She found that if she went into the train toilet and carefully put make-up on before the train went underground, she could bear the anxiety far better, as though the make-up could somehow protect her if the tunnel collapsed. While she understood consciously that the link was untenable, it nonetheless enabled her to keep travelling.

Of course, none of these defences are, in themselves, necessarily all bad, but they become a problem when the fantasy solution is opted for so enthusiastically that fashion and cars start clogging up the planet, causing people to feel anxious, and then to vote for yet more of the sorts of leaders who promise an endless supply of fashion and cars. In this case, our self-soothing immortality system begins to put us at risk.

Another component of our placatory self-deceit comes in the form of what Becker terms "transference". In psychoanalysis this, *very* loosely speaking, means treating your therapist like a parent, i.e. a big, powerful figure who protects and provides for you at the same time as pissing you off. For Becker, in this context, it means we look for potent, larger-than-life figures to latch on to – maybe politicians or pop stars, youthful tech billionaires or teachers, handsome faces on Tinder, or anyone, absolutely anyone, who seems special and brilliant, and who appears to us to imbue life with sense and direction. An identification with these figures can be so powerful, and appear so promising, that we might find ourselves, say, smoking because cool people smoke – because coolness has an enduring, timeless quality that promises more than simple good health. If you're healthy you still die at some point, but if you're Humphrey Bogart you are immortal. Like buying a cashmere jumper in response to a flood, it makes no sense, but that doesn't stop people doing it.

It appears that Becker's field of interest developed, at least partially, out of his experiences as a Jewish soldier in the American Army, liberating prisoners from death camps at the end of the Second World War. He witnessed directly some of the worst crime scenes imaginable. How was it possible that human beings could inflict these levels of suffering on one another? What was going on in their minds to make these levels of denial possible? His theory attempts to have something to say about this, explaining why people might turn to charismatic leaders in moments of widespread, life-threatening poverty. Rather than hatching plans around fairer wealth-distribution (which many rich, powerful people hate and actively discourage, of course) it might seem easier to invest in the figure of a father-of-the-nation-type "Big Daddy", who promises to make things better by protecting the interests of your religious/ethnic/racial group by getting rid of a load of other, slightly different people who also need food, warmth and space. Becker thus considered our passion for the denial of death to be a great source of evil. If reminders of death make us grab at worldview-supporting ideas and narratives – and the shiny figures that also support them – they risk pushing us toward cruel, idiotic and self-esteem-based solutions that simply don't work. In relation to the climate crisis, this might mean that people would be inclined to reach for SUVs, luxury handbags, trips to expensive resorts, politicians who surround themselves with the blatant signifiers of outrageous fortune (including having younger, much-better-looking partners) or any number of potentially climate-damaging activities, in order to give themselves the sense that nothing could possibly be wrong: "Look around you – life's great! You're cool, so why worry?"

This is made all the more intense, the more technologized and risk-averse we become. It's as if we actually start to believe that death has very little to do with us any more – it's just another scientifically solvable problem that only affects "other people". Becker's ultimate idea was that people needed to wean themselves off these pacifying myths in order to pave the way to ask difficult questions about life, death and the ethics of coexistence.

> "I can't even get angry about it anymore. I used to, but it's just too tiring. I don't get why some people care and some don't – why some people don't even really seem to know about it. They just carry on screwing things up for everyone, and they don't even realize. It's too weird!"
>
> Barnaby, student

Green Terror

For Janis L. Dickinson, a lab-dwelling scientist, to risk assertive predictions about climate change in full view of her employers and students, an evidence-based framework might be expected to support her thinking. In her essay "The People Paradox: Self-Esteem Striving, Immortality Ideologies, and Human Response to Climate Change" (2009),[15] Dickinson uses Terror Management Theory – a development of Becker's work – to elucidate ideas around climate denial, and to offer some idiosyncratic ideas for combating it.

In the 1990s, a group of social psychologists set out to test Becker's theories – and to demonstrate how reminders of death might impact our behaviour. Under the banner of Terror Management Theory (TMT), hundreds of experiments were conducted showing how people's thinking might be affected by "mortality salience" (TMT's preferred term for the awareness of death). For example, if you asked people to consider their own death, how might it affect their subsequent questionnaire answers?

Not everyone was influenced in the same way, of course, but, as it turned out, it seemed enough people could be swayed in a predictable direction to have a notable impact on, say, voting or shopping choices. Still, it wasn't as simple as firing off a reminder and getting a result: different reminders generated different responses. One key distinction was between conscious and non-conscious prompts. A conscious reminder – like being shown a film of someone dying – was more likely to generate a "proximal defence", i.e. a set of rationalizations. So, in relation to the climate, this might involve questioning the science, denying that human beings have anything to do with it, or pitching it so far into the future that you didn't care either way. In 2006, a survey conducted by the University of East Anglia found that watching the climate disaster movie *The Day After Tomorrow* actually made people *less* likely to believe in the risk of extreme weather events.

Then you had to contend with non-conscious reminders. A non-conscious reminder could be something like a subliminal word-flash or even, apparently, a photo of a humanoid robot. This type of reminder was more likely to trigger a "distal defence", meaning an idea or behaviour that reinforces self-esteem in response to a threat – like putting on make-up in the toilet of the train you believe will kill you. (The terms "proximal" and "distal" come from anatomy and are used to designate nearness or farness from the torso.) The thing about non-conscious reminders is that they don't trigger the full, unsettling effect of a conscious brush with death – you don't even realize it's happened. Therefore you don't have to go into overdrive, reasoning the problem out of existence. Instead you might find yourself mysteriously drawn to worldview-supporting ideas, without noting any link between the initial noxious reminder and the veiled rebuttal. In one experiment, people were given a non-conscious death prompt and then asked to choose between two politicians – one who stressed national pride and achievement, while the other promised to listen to citizens. Subjects in a state of non-conscious "mortality salience" were more likely to choose the nationalist. This is surely something that we have seen come into play in global world politics over the past few years.

It could also be the case that distal defences were aroused by a conscious death prompt. For instance, people who had just watched frightening news items were more likely to respond to luxury advertising, seemingly upholding Becker's theory that people are inclined to invest in their symbolic selves rather than in their mortal, corporeal selves when faced with ideas around extinction. And, furthermore, that this misguided tendency would push us, as individuals and as a species, toward serious jeopardy.

For the Birds

Not only does Dickinson show remarkable acuity about the future, she offers some brilliantly inventive and unorthodox ideas about possible responses to the climate crisis. If you're sick of hearing all the usual stuff about giving up beef and challenging the fuel industries – both obviously good ideas, but how often can they be repeated before they start to blur into the background? – it can be inspiring and refreshing to come across something that involves a little more lateral thinking. Dickinson's first idea involves redirecting the power and energy of urban gangs into conservation. Here, she cites a journal article by José M. López from 1991 that looks at three books focusing on the value systems of gangs, in particular the idea that they are cultures of honour that offer greater prospects of self-esteem for members than either regular schools or jobs.[16] Dickinson's argument is that environmental projects can offer marginalized young people an alternative context within which to establish self-esteem. Like gangs, whose names, codes and insignia might seem to transcend life itself, conservation groups can also be "immortality projects" – albeit ones where you're also less likely to achieve immortality by getting killed.

This isn't simply some hopeful ornithologist's fantasy, but has actually been tested across the US, where youth conservation corps programmes apparently had more applications than they could accommodate, especially from poorer inner-city youths. Dickinson acknowledges that simply redirecting potential gang members toward conservation might just create new gangs, but that "outgroup antagonism", or gang warfare, could be avoided if young conservationists were given the opportunity to address politicians and the wider community, giving them greater social capital, thereby circumventing their need to attain it through crime.

Dickinson's second idea involves harnessing what she calls the "charisma of birds". I should admit that I hesitated slightly when I first read about this, but it's perhaps the idea that's made the most difference to me during the writing of this book. I know it doesn't have the immediacy of, say, going plastic-free or ditching your car, but after reading about it, I immediately installed a bird feeder outside the window of my consulting room, just to test it out. Since then I have been visited constantly by a stream of coal tits, blue tits and robins, and it never, ever ceases to be astounding and mesmerising. The people who come to see me for therapy also seem to like the birds, and often go off and get their own bird feeders. There is something completely astonishing and adorable about these tiny, colourful, airborne creatures. I challenge anyone with a window-mounted bird feeder not to become a diehard environmentalist. Seeing birds close up is a daily reminder of the beauty, fragility and cheekiness of nature. It also seems to have a positive impact on human relations – I *love* seeing other people enjoying "my" birds. (I haven't given them names yet, but I'm getting to know their ways. Coal tits love chucking seeds out of the feeder, while robins are leisurely eaters.)

Dickinson argues that birds can become transference objects in place of film stars or romantic crushes. Flight is mind-boggling and birds are really good at it; therefore they are liable to captivate us if only we will let them. Dickinson tells us: "When it comes to climate change, birds may be superior archetypes to other charismatic organisms. Not quite celestial, they have the unusual capacity to take to the sky with a beauty, mystery and charisma that renders them elusive, godlike and apart from us. These characteristics make them ideal symbolic 'transference objects' on which to project a striving for immortality."[17] So instead of imagining that a parental figure will save you, you can get into birdwatching, which is a culture of honour and a competitive sport with opportunities for self-esteem, plus "the idealization of birds may have anxiety-buffering effects". I feel bad that I initially doubted this idea even for a second, but perhaps it's sometimes difficult to register brilliance when you first see it. I can absolutely testify to the idea that birds make people feel better. In fact, I think I might try to get my coal tits registered as official therapy animals.

Still, no solution is perfect. As Sally Weintrobe observes in *Engaging with Climate Change* (2012), even so-called nature-lovers can be inclined to exploit and consume the animals and landscapes they purport to adore: "A person who loves birds and also loves walking [...] may be seduced away from loving birds into seeing him/herself as the owner of the most special binoculars, the best new walking poles, and with a superior capacity to find the best and most exotic global locations in which to walk and see birds. The result is an increased carbon footprint which will adversely affect birds as well as the people themselves."[18] As I type this I realize that my bird feeder is made from Perspex and was delivered by Amazon. I'm ashamed – but I guess this goes to show that you can never be perfect, even when you think you're giving it your best shot. If you buy a bird feeder, please learn from my mistakes!

As well as being elegant and original, Dickinson's suggestions are underpinned by the helpful idea that simply hectoring people to make sacrifices, change their behaviours and take responsibility for their environment may alienate them further. This is the nature of the "people paradox" in the title of her paper: if you actually manage to persuade people that the climate emergency is real, you may push them deeper into denial, heightening the polarization between your respective positions. Therefore, perhaps, you need to coax people into valuing the planet and doing whatever they can to protect their immediate surroundings, and to foster a connection to their local environment by showing them how pleasurable and satisfying this can be. Some might say that the time for this sort of nicety has run out, and that it's far better to make like Greta and try to get people to panic, but perhaps all strategies need to be kept in mind at all times. And you could also say that keeping the awe-inspiring beauty of birds in mind, while panicking somewhat, is possibly the best outcome of all.

Which brings us on to one last idea from Dickinson's paper: the notion that you don't necessarily have to change people's minds that much if you're clever about which bits you change. She refers to *The Creation: An Appeal to Save Life on Earth* by Edward O. Wilson (2006), in which the scientist attempts to persuade American "biblical literalists" that conservation is a better idea than bringing on the apocalypse. Wilson argues that you can reconcile a respect for the Book of Genesis with a concern for biodiversity – God put humans in the Garden of Eden, telling them to guard and protect it, after all. So scientists and evangelical Christians needn't be at odds with one another, in spite of their differences.

Dickinson reminds us that holding contradictory ideas in mind is very human, and that people don't need to be coerced into being 100 percent consistent. If you actually care about saving the planet, rather than being "right", you might be better off letting people combine a concern for the planet with their existing worldviews, rather than going at them Richard Dawkins style, haughtily dismissing their belief systems and demanding that they change. This also goes for lefties like me, who think capitalism might have something to do with the climate problem. (Radical!) This shouldn't mean I spend all my time trying to overthrow capitalism, meanwhile dismissing the idea that there's any point in other forms of environmental action. I can be a green consumer while I wait for the revolution. (Ugh … *checks watch* …) And if it turns out that I'm completely wrong, and that the "green capitalists" are right – "Keep spending, Elon's on it!" – I'll be delighted to admit ideological defeat, at least with regard to the climate.

You've Been Reframed

Changing people's minds *just a little bit* is a major new area of research, made famous in 2016 by US politician Bernie Sanders and his army of door-to-door persuaders, who called on the latest developments in psychology and neuroscience to try to win round right-wing voters. Legions of Democrat supporters were given "persuasiveness training", enabling them to have civil conversations on the doorsteps of people whose opinions sat at the very opposite end of the political spectrum. So how does it work?

Current research on persuasiveness has its roots in classic twentieth-century questions about human beings' capacity to change their behaviours, and their minds. Are our actions, thought patterns and emotional responses permanently hardwired, or can they be altered? Psychoanalysts, psychologists, philosophers, chemists and neuroscientists have approached these questions in various conflicting and contradictory ways, coming up with a vast range of mutually exclusive answers: it's all down to the brain, the mind, society, the individual; change is easy, change is impossible; drugs are the answer, or surgery; or mindfulness, or brain training, and so on. The questions can seem utterly unanswerable. However, "mind changing" is now a well-funded area of research, due to the fact that it's extremely important to people interested in money and power. Businesses and political parties famously use data mining in order to manipulate us, and who's to say where things might be headed next? While to some, mass drugging seems a long way off, to others it's already widespread. What if you could make whole societies more complacent simply by making them "happier"? The *Journal of the American Medical Association* reported that in 2012, 13 percent of the US population were on SSRIs (see page 16), and the numbers continue to rise. Alongside this, Americans' addiction to painkillers has reached epidemic levels, to the point where more people die from opioid overdoses than car crashes (130 a day in 2016 to 2017).[19]

If science can provide us with more information about what's involved in human change, how can it, or will it, be used? Happily, at least for now, neuroscientists themselves are keen to remind us that human minds are extremely complex, and that their field is relatively undeveloped; a rudimentary understanding of the biology that underlies thoughts and feelings hardly gives you magic powers. Plus, if you make changes to one thing, you risk unwanted knock-on effects elsewhere; change is hard to control. For example, oxytocin is a hormone that acts as a neurotransmitter; it's linked with empathy and bonding. Therefore, one might think, it could be a good idea to increase oxytocin levels in people who lack empathy – sociopaths perhaps – thereby making the world a better, safer place. The problem is that increased empathy appears to be linked with increased aggression, which is in turn associated with high levels of vasopressin, another hormone. This phenomenon has been closely observed in prairie voles, which are famous for mating for life. In mating season, as the dear little voles feel intense attachment to their mates, they become intensely aggressive toward any other sexy voles who might threaten their marital bliss.

The more you love one person or group of people, the more likely you are to hate another it seems. As researchers at University of Buffalo, New York, discovered, if your sympathy for one person is aroused, you will be far more likely to act in a hostile way toward a third party. The Greek word for empathy was also used to mean something like bias or partiality, and it appears that this connection can now be scientifically backed up. Because of this, it's sometimes suggested that internally friendly climate groups who manage to crystallize hatred around named polluters (the CEOs of ExxonMobil or BP, for example) actually provide quite a structurally sound model for bringing about change. This, however, is in direct contradiction to the philosophy of groups such as Extinction Rebellion, who insist on non-hateful attitudes towards individuals – even those whose activities you abhor.

Politicians and military leaders have, of course, understood this dynamic for centuries, and attempted to use it to gain advantage. This is the problem with so many neuroscientific "discoveries": they simply give us a mechanistic explanation of something we already know. Still, recent studies of brain reactions promise to throw new light on the thorny issue of people's political allegiances. While, on the one hand, we are told that elections can be won or lost on Facebook's dark ads, scientists at Virginia Tech, Roanoke apparently have evidence to suggest that left- and right-wing brains are fundamentally different. This may be relevant to discussions around climate change, as those on the political right, such as Trump, Scott Morrison and Jair Bolsonaro, have typically been associated with climate denial, while climate concern itself has often been attacked as a leftist conspiracy. Anyhow, it seems you can't just be persuaded to swing from one side of the political spectrum to the other at the whim of a Slovenian hacker. "In fact, the responses in the brain are so strong that we can predict with 95 percent accuracy where you'll fall on the liberal–conservative spectrum by showing you just one picture," claims Read Montague, the research leader.[20] The picture in question is a photograph of a badly mangled animal. People whose brains register a particularly strong disgust response are overwhelmingly more likely to have right-wing political leanings. (Left-wingers may be upset or unsettled by the images, but not disgusted.) This may also show itself as a particular concern with purity, cleanliness and the appearance of consistency. Anxieties about multi-racial societies, differing sexualities and abortion, it seems, are experienced by certain people as threats to the integrity of the social corpus. It's interesting to see how this research intersects with the findings of TMT, perhaps throwing light on the fact that some people's choices were much more affected by mortality salience than others.

Not only do the brains of conservatives and lefties "light up" in different ways, they even appear to be structurally different.

Left-leaning people apparently tend to have a more developed anterior cingulate gyrus, the part of the brain that processes new information and is involved in making choices, while right-leaning people are liable to have a larger right amygdala, a deeper, more primitive part of the brain that's involved in processing emotions, perhaps most notably the feeling of fear. (Research in this field is relatively new, so it's as yet unknown whether the shifts in political allegiance that may take place over a lifetime are accompanied by changes in the brain.) In a sense, you could say it's yet another example of neuroscientists backing up a familiar cliché: that right-wing people are instinctively scared of "something out there", while left-wing people tend to have a more analytical approach to problems (which leads us to come up with idealistic but impracticable schemes – just to demonstrate that I *can* listen to other points of view). But what are we supposed to do about it?

Some have suggested that this sort of research might lead toward greater tolerance on both sides: if one's political opinions are simply a sublime and unavoidable biological matter, then there's little point in getting into a fight over it. And there's also the fact that people have been known to switch their political allegiances when fear has got the better of them, or when they were encouraged toward a more analytical approach. (If this sounds slyly biased in favour of left-wing politics, it's interesting to note that Britain's ultra-right-wing *Daily Mail* carried an article stating that "right-wingers are less intelligent than left-wingers",[21] citing research published in *Psychological Science* in 2016. The problem is that this greater "intelligence" also apparently gives them a greater capacity to change their minds – although only in response to persuasive, credible arguments – while right-wing people are apparently notoriously ... well ... conservative.)

If physically altering a person's brain seems a lot to ask of a political argument, there are softer ways of pushing people around. Matthew Feinberg (a professor of organizational behaviour) and Robb Willer (a sociologist) produce work on the art of persuasion, suggesting that one must "reframe" or dress up one's arguments in the clothing of the other side. If you want to persuade climate deniers to reduce their meat consumption, you might be wise to tell them that vegetarians are less susceptible to colon cancer. The beauty of it is that no one is expected to change their mind too much. Food producers are already well onto this and are moving away from the words "vegan" or "vegetarian" on their packaging, as this apparently alienates meat-eating customers. Instead there is an emphasis on tastiness – words such as "sizzling", "sweet and smoky" or "succulent" are popular – and the use of positive terms such as "plant-based proteins", as opposed to negative terms emphasizing an absence, such as "dairy free". This way, they try to include meat-eaters in the set of people who may enjoy their product, rather than suggesting that it's aimed exclusively at vegetarians or vegans.

Way before the advent of contemporary psychology, various religious practices aimed to encourage people to shift their thought patterns and redirect their impulses. Meditation or prayer could supposedly help soothe the mind and channel energy into more serviceable activities. Now, brain scans tell us that these ancient habits actually work. It doesn't happen instantly, but careful, laborious rethinking can bring about chemical change. This is good news for psychotherapists, who've long argued for the benefits of a gradual, thoughtful approach to psychic transformation. And it may also be good news for politicians – the question is, *which* politicians? If, loosely speaking, the two main political tactics are either to frighten people or to explain things to them, which strategy will prove more effective over time? If some brains supposedly respond better to the former and some to the latter, what happens when you start muddying the waters with tactics such as "moral reframing"? Not to mention dark advertising and Prozac. As Lady Hale, the outgoing president of the UK's Supreme Court, said, "Everyone is persuadable."[22]

This has to be good news for those of us who are tormented by the looming climate crisis. Not only may we be able to settle our own minds a little by using therapy, meditation, relaxation techniques or whatever works best for us, but we can realistically hope that those who are currently in denial may be brought round. If the main thing that keeps people from addressing the problem is a wish for things to remain the same, it will surely soon become apparent that the best way to prevent catastrophe is to take active steps to address it. In a sense, climate deniers and climate activists want much the same thing: to keep living life in the best possible way. For those in denial, perhaps, it will be their own conservatism that wakes them up to problem: if we want to preserve our ways of living, we'll have to act decisively to make that possible. If we keep consuming at the current rate, collapse is inevitable. So if you have parents, aunties, bosses or partners who seem to have their heads stuck in the sand, it may not take much for them to make the shift from disavowal to decisive action. If what they want is to carry on with the nearest thing to life as they know it, they'd be wise to get real about the sorts of sacrifices that might entail. Who knows, the most chronic naysayers may turn into the most passionate conservationists, as they are the ones who have the biggest stake in forestalling change. So be nice to your complacent friends and relatives – keep them on side and charm them, because when it comes to the crunch, they hold the key to planetary survival.

The Power of Friendliness

To go back to the wisdom of Edward O. Wilson, what if it's not simply a question of left vs. right, science vs. religion, eco-warrior vs. climate denier? To believe in these binaries is to uphold Janis L. Dickinson's tragic prediction. An antidote to all this is the work of the Californian psychoanalyst and philosopher Donna M. Orange, whose book *Climate Crisis, Psychoanalysis, and Radical Ethics* (2016) is a hopeful appeal for a kind of extreme hospitality. If Becker dates the origins of denial to human pre-history, Orange is more focused on the forms of denial that made slavery and colonialism possible. Sally Weintrobe, of the excellent birdwatching admonition, also writes about "colonization of the mind" and "splitting" – a form of psychic defence that might just cause you to buy a mass-produced plastic bird feeder from a controversial and much criticized multinational while writing a book about climate change.

In post-industrial parts of the world we live in societies made wealthy by the exploitation of others. More than that, there are reminders of it everywhere: statues, galleries, books, public holidays, all pointing us back to the crimes of the not-so-distant past. For people to have committed these atrocities – in broad daylight and under the protection of law – there must have been a passion for denial that seems outlandish, until you remember that nature is currently being polluted at a rate that's putting all living organisms in danger; not to mention the ubiquity of poverty, famine, preventable disease and the countless civilian casualties of war. For Orange, the climate crisis corresponds directly with the cruelty of our recent history.

Our current capacity for denial is part of a wilful project of pushing blatant inhumanities out of our minds in the hope of maintaining an apparently stable status quo. While we would no longer feel OK about keeping a slave at home, we may in blissful – or wilful – ignorance buy objects manufactured using enforced child labour. In order for us to wake up fully to the climate crisis, we need to open our eyes and admit to the horror of our history, and our present, and then to make it our urgent business to do all we can to make reparations. This means everything from giving up unnecessary travel (which is scary as we have been conditioned to believe we "deserve" our short hops abroad for a weekend of merriment), to opening our minds and our spaces to the people who have been disenfranchised by war, climate change, international systems of debt, global outsourcing and its related labour rights abuses.

Weirdly, Donna Orange doesn't seem overly bothered about human beings' systematic abuse of animals, and treats vegetarianism as a side issue. Still, I suppose it's forgivable in light of the rest of her work. For her, it's vegetarianism; for me, plastic bird feeders: all of us have our blind spots. Perhaps somewhat idealistically, she thinks therapists have a slight head start when it comes to politically motivated hospitality due to the fact that many of us already invite all sorts of people into our houses, and, if we are at all decent, put a generous portion of our practice aside for those who pay very little. Reading Donna Orange's entreaties to put other people's needs ahead of personal comfort, and to believe in the possibility of a kind of wild generosity, doesn't seem so crazy in the light of the sorts of things so many people already get up to in their ordinary working lives. Teachers, nurses, doctors, firefighters and social workers put their radical ethics into practice on a daily basis. If anything good comes out of the coronavirus pandemic, perhaps it will be to make the world value the work they do much more, and to notice that they – alongside other essential "low-skilled" members of the workforce: childminders, carers, grocery store workers – are far more essential to a habitable society than CEOs, university deans and hedge fund managers.

Basically, it's not *that* outlandish to imagine a situation where people put aside their differences and pull together. In fact, you could even say that this is one of the most standardized forms of human behaviour. We didn't get this far without a certain degree of fellow-feeling – even if it has traditionally gone hand-in-hand with a disturbing capacity for violence and denial. But who's to say we're not ready to take things to the next level and see what it's like to *really* begin to take coexistence seriously? In the next chapter, we'll have a closer look at some of the personal qualities that might help to make this possible.

DENIAL: CHAPTER OVERVIEW

1. Acknowledge that all of us are bound to practise some form of denial – human culture is mainly about this. In fact, "functional denial" may be an essential part of day-to-day survival.
2. If you can be honest about your own capacity for denial, it will probably help you to understand that of other people.
3. Understanding and empathy are useful in persuading other people to be kinder to the planet. Try not to think of climate deniers as the enemy: they are just people who have yet to get the message.
4. Change isn't straightforward and some people's resistances may be extremely deep rooted and seemingly inaccessible. This isn't because they're stupid or stubborn but because, like all of us, they are beholden to feelings and attitudes that seem entirely natural and inevitable to them, even if these may seem weird or obnoxious from the outside.
5. Don't give up on people whose ideas about climate change seem uninformed or even wilfully ignorant. If you can find a subtle argument that fits with their core beliefs while demanding that they make behavioural changes, you've won. In a nice way.
6. Try to exercise judgement over when to harangue and when to inveigle. Oscillate at random to keep them on their toes.
7. Get a bird feeder – I promise it will change your life, not to mention the lives of everyone who visits you!
8. If the future demands insane levels of generosity from people, maybe it could just work. We have to keep believing that, even in the face of disappointment.

5

FRIENDLINESS GETS RADICAL: DRAW ON YOUR INNER RESOURCES

"We have today the knowledge and the tools to look at the whole earth, to look at everybody on it, to look at its resources, to look at the state of our technology, and to begin to deal with the whole problem. I think that the tenderness that lies in seeing the earth as small and lovely and blue is probably one of the most valuable things that we have now."

Margaret Mead, cultural anthropologist,
Earth Day speech, 1970

If the climate crisis has been caused by the overuse of material resources, what are the *immaterial* resources we can draw on to begin to turn things around? In this chapter we will focus on potentially limitless resources such as communication, curiosity, generosity, social engagement, ingenuity and honesty, to see how we might be able to help ourselves and others to feel better without blotting out uncomfortable truths. Of course, some of these resources also rely on another resource – time – that's extremely scarce in many people's lives. This can lead to a situation where you feel you're never doing as much as you should, which leads to guilt, which leads to helplessness, which leads to misery. So the point here will be to try to be realistic about the resources at your disposal, and to stay cognizant of the fact that everybody is different. As long as you're doing whatever you can, without wiping yourself out in the process, you're on the right track. (Unless wiping yourself out, saint-style, is your particular cup of tea, in which case keep going – just try to do it as resourcefully as you can.) It may even be the case that you can make small shifts in your attitude and outlook that take no time at all, but that are then right there, ready to go, as soon as you run into a situation that calls for them.

Before looking at a few activities and attitudes that might be helpful, perhaps it's worth taking a glance at the larger powers that can get in the way of constructive environmental action. It's extraordinary and terrifying that people have been talking about the threat of ecological breakdown for centuries, yet so little has been done to prevent it. If we have some kind of overview of the history of the environmental movement, we might see more clearly some of the forces that militate against it. In this way, we can be wiser to them when they get to work on us. If we're going to make use of all the precious immaterial resources at our disposal we need to know about the counter-movements that will attempt to exhaust, discredit and ignore us, so we can target our energies as efficiently as possible.

Early Environmentalists

Alexander von Humboldt, an early nineteenth-century explorer and scientist, is sometimes referred to as the first environmentalist. Without meaning to diminish his extraordinary achievements, this is a load of colonialist nonsense. Still, he did manage to whitesplain a great number of things already perfectly well understood by countless indigenous communities, underpinning his advice on looking after the planet using the methods of Western science. In this sense, he's an excellent person in that he can speak the language of power in order to explain what the people in power are doing wrong. An independently wealthy polymath, in 1899 he travelled to the parts of the Americas colonized by the Spanish in order to study the environment and its inhabitants. He spent five years collecting samples and evidence before coming back to Europe and devoting the rest of his life to documenting and interpreting what he found. His big take-home message was that everything is interconnected and that human activities have a tendency to make a mess of the environment by interfering with its complex, delicate systems. To his very great credit, he saw that this was not only the case for plants and animals, but also for colonized peoples who were robbed of their histories and traditions. Von Humboldt was a great influence on Charles Darwin and Alfred Russel Wallace, the founders of modern natural history. So basically, there's no David Attenborough without von Humboldt; his influence on environmental science has been enormous.

After von Humboldt, "conservation" became a serious preoccupation throughout the nineteenth century, with legislation protecting everything from birds to forests to air and water from the encroachments of industrialization. Unfortunately for us, the dominant idea seems to have been to sustain capitalism and colonialism while taking care of this and that, so you can keep extracting while feeling good about yourself.

Following on from Romantic thinkers such as Rousseau and Wordsworth, later nineteenth-century artists and writers such as John Ruskin and William Morris took a more holistic view of environmentalism, suggesting that people needed to radically rethink their ways of living and to get more in tune with nature. Because these people realized that bitty changes in legislation weren't enough, and that a major structural rethink was required, they were relegated to the category of charming cranks. It was fine for them to be weird and arty – it was even a nice reminder that other ways of being might be possible – but that didn't mean you actually needed to do anything about it. Dreamers are the conscience of any society, but the risk seems to be that their little taps on the shoulder are too easy to subsume. You could put up some lovely William Morris wallpaper and get on with running your coal mine, keeping it as far away from your well-managed garden as possible.

The early twentieth century, with its two world wars, sped up the destructive power of technology. Using the logic of emergency, people could ignore the niceties of the nineteenth-century conservationists and start developing weapons, materials and farming systems that took trashing the planet to previously unimaginable levels.

After the Second World War, the marine biologist Rachel Carson began to study the effects of pesticides – particularly DDT (or dichlorodiphenyltrichloroethane) – on the environment. In 1962, she published her seminal book, *Silent Spring*, detailing the effects of synthetic pesticides on biodiversity, and exposing the greed and dishonesty of the chemical industry. Even before publication, her publishers knew she would be under attack from the people who had a vested interest in gagging her, but perhaps even they had no idea how bad the attacks would be. Modern readers, more versed in the forms of misogyny that clutter up the internet, might not be surprised to hear that she was accused of everything from not knowing her subject, to putting people at risk of malaria, to being a communist (shock, horror). Thankfully, the smear campaign didn't work. Her book is a classic, is still in print, and is considered one of the best science books of all time. Not only that, but the use of DDT has been phased out, and Carson's work led to President Nixon setting up the Environmental Protection Agency in 1970.

Having said that, it's shocking to read some of the work of American environmentalists from that era and to realize how little their urgent warnings have been heeded. On 22 April 1970, America hosted the first Earth Day, a huge event, in which 20 million people participated. Out of this came a book, *Earth Day – The Beginning*, with essays by politicians, writers and activists. One of the most shocking aspects of this publication is that it could have been written yesterday. On the very first page, it speaks about "a bizarre alliance that spans the ideological spectrum" and acknowledges the links between pollution, war and racism. It talks about "our rush toward extinction" and the urgency to act before it's too late. Inside is a short play by the social worker Freddie Mae Brown, who calls for action at the same time as explaining why marginalized people in industrialized countries might not see it as their most pressing issue, but "more of a white mothers' club thing". (Pretty much the same accusation that has more recently been levelled at Extinction Rebellion.)

So that takes us up to the present, where we can see perfectly clearly that environmentalists are still being ignored and bullied by people who value money more than they value life. But how can we prevent that from happening?

Greta and XR, You're Probably Our Best Hope

In the twenty-first century, it's perhaps become easier to see why and how environmentalism gets stuck. It isn't because ordinary people are ignorant or don't care (e.g. the insulting myth of the feckless poor who don't recycle because they don't know any better) but because it's not in the interests of the super-rich. Capitalism, as people have been trying to tell us for quite some time, is unsustainable. In the words of the environmental activist and writer George Monbiot, "Perpetual growth on a finite planet leads inexorably to environmental calamity [...] There must always be an extraction zone – from which materials are taken without full payment – and a disposal zone, where costs are dumped in the form of waste and pollution."[23] You simply can't go on expanding forever. The problem is that most of us have been persuaded that there's absolutely no alternative, or that the only alternative is to recreate Soviet Russia or Mao's China, and obviously that's not popular.

What else can we do? The first thing seems to be to get large enough numbers of people to acknowledge that the climate emergency and ecological breakdown are real, and then to mobilize them to act on that knowledge. The environmental activist organization Extinction Rebellion (abbreviated as XR) set about doing this in a very purposeful, calculated way, studying the workings of successful protest movements, from American civil rights to India's fight for independence, to the British overthrowing of the poll tax. As Roger Hallam, one of the co-founders of XR, puts it in the book *This is Not a Drill*, "We went to the library. We studied decades of work looking at organizational systems, collaborative working styles, momentum-driven organizing and direct-action campaigning."[24] From here, they concluded that non-violent action was the way forward, following what they call the "civil resistance model". This involves getting large numbers of people to break the law in cities, maybe by blocking roads or sticking themselves to buildings. The law-breaking must be non-violent, it has to go on for at least a few days, and the whole thing must be enjoyable for the people involved (in order to bump up numbers). This way, you exert pressure by costing governments money, at the same time as making them look bad if they suppress you. In turn, you open up a crack in which something can be expressed or demanded in order to bring about change.

International uprisings in November 2018 and April 2019 succeeded in enabling people all over the world to get out and cause peaceful trouble in the name of protecting the planet. In October 2019, another wave of actions took place in London and elsewhere, but this time there was something of a backlash, particularly after a protester was beaten by commuters who were delayed on their journey to work as he stood on top of their train. Not only did people question the tactic of targeting public transport – the Tube is one of the greener ways to get around London, and less well-off people rely on it – but there were also criticisms of XR's exhibition of white privilege in their policy of inviting arrest, which could be seen to exclude people of colour who might not be so inclined to expect fair treatment from the police. In spite of their sincere soul-searching in the face of all this, the seemingly inviolable movement took a hit. Here, some of the most devastating criticisms came from people who were largely in the same camp, climate-wise at least, and perhaps for this reason were particularly unsettling. There's nothing worse than being bashed by people you basically agree with. When, a few months later, XR were listed as an extremist movement by the British police – alongside neo-Nazis and other groups promoting extreme violence – it was possibly easier to deal with. The police immediately apologized and retracted, and the overall result may have been renewed sympathy for a collective of people who are quite clearly doing their best to bring about changes that all of us stand to benefit from.

Alongside XR, but quite independent, came Greta Thunberg, the lone Swedish teenager who inspired millions of schoolchildren to leave their classrooms on Fridays and strike for the climate. Perhaps because of her youth, passion and incredible capacity for straight talk, she has so far been singularly admired by all but the shadiest climate deniers and a few weird outliers who either haven't yet grasped how serious things are, or who wouldn't recognize a good thing if it appeared right under their nose. As yet there's no good reason I know not to get right behind Greta and start fighting for change. The only very minor criticism, which certainly isn't of Greta herself, is that World Economic Forum/UN-type people still seem able to cheer her on, imagining their virtue signalling is, in itself, enough, and it doesn't then necessarily follow that they need to do something about it. Anyhow, she's onto them and I'm sure she has it in her to pierce their defences eventually.

So, it isn't hard to see how environmental movements have a truly Sisyphean task in keeping people on side and keeping up the pressure. There are powerful forces working against you, and it's utterly exhausting. But there are qualities that are free to acquire and potentially inexhaustible. This next part might sound a bit sappy, but I hope it'll become apparent that there's a point. While there are obvious things you can do, like stop flying and start writing letters to MPs and/or gluing yourself to buildings (see Chapter 8), there are also a number of traits, skills or qualities you can aim to acquire, and then apply to *everything* you do, not only things that are directly climate related. *Planet-love can be an overall attitude, not an extra set of gruelling tasks.* Here's a checklist of potentially limitless resources that are likely to make life better for you, the climate and everyone you encounter.

Communication

As outlined in the previous chapter, there are ways of communicating that are more likely to get your point across, which isn't to say that all communications on the subject of the environment should be carefully controlled and planned with precision. There also have to be times for authentic, surprising discussions, where you allow yourself to say difficult things, and to listen to other people doing the same. In my work, I'm sometimes asked, "What's the point in talking about things if the person you're telling it to can't actually do anything about it?" While I see the logic, I don't agree. You can *make* it feel pointless if your aim is to browbeat the other person into feeling as angry and impotent as you do, but this is thankfully quite a rare conversational style. (And therapists are trained to work with it by not taking it at face value, but to try to discover why the person chooses to behave in this way toward their interlocutor.) Most people find that putting feelings into words, and formulating ideas around those feelings, can make a huge difference to their mood. And that's even before the other person says anything brilliantly insightful back.

There are few things worse than feeling trapped in your skull – especially if you fear that your thoughts and feelings somehow set you apart from the rest of humankind. Rather than frightening people away by expressing yourself, you're more likely to make them feel better in the knowledge that others are as weird/sad/troubled as they are. (But if you're convinced that the stuff you need to talk about is totally socially unacceptable, maybe try it out on a therapist first.)

Being able to bare your soul, and to let others do it back to you, is a skill that will protect you – and other people – from loneliness. The climate is everyone's problem, so don't let your anguish isolate you. Addressing and being addressed are vital to life: babies need to be responded to constantly if their brains are to develop in ways that enable them to get along in the world. Communication isn't an extra, it's a necessity. So keep talking, and maybe have a go at writing too.

Writing is great because you can try ideas out in secret before testing them in more public spaces, plus it can help you to formulate thoughts. You can scribble things in private notebooks or post them online. You can write a certain number of words per day, or keep going for a certain amount of time, or you can just do it when a build-up of thoughts is driving you nuts. Writing can help you to know your own thoughts better, which makes them easier to share with others. Whether you keep it to yourself, share it in the form of poetry, journalism or Twitter posts, or direct it toward MPs and CEOs, writing can be a major tool in feeling better about the state of the planet.

Curiosity

In order to like life, and to have thoughts to exchange with other people, it's important to stay curious. Thinking and wondering stops you sinking back into yourself – although even introspection can be strangely edifying. Your mind is the object of study you have most unlimited access to, and knowing one mind well can help you to think about others. Focusing on yourself can sometimes feel self-preoccupied and pointless, but you can use it as a springboard to thinking and caring about other people. Being outwardly focused militates against anxiety and depression by fishing you out of the swamp of uncomfortable thoughts and feelings. However, knowing yourself better can stop you falling into mental – and interpersonal – traps. Turning your curiosity both inwards and outwards helps you to construct a lively place from which to think.

Finding out as much as you can about the way the world works, from ecology to geology to psychology to politics, gives you a much greater sense of contact with your surroundings, and beyond. How you do it is up to you. You don't have to read an encyclopaedia (as if anyone ever did that anyway); you can watch documentaries or YouTube videos, read the papers, look at social media, listen to people, read fiction, watch movies – while keeping in mind the fact that the threat of the echo chamber is real, and that stepping outside your regular circuits will help you to see things better. I just subscribed to the *Spectator USA*. WTF. It actually has some great articles. Immersing yourself in the world is clearly a great way to stay in touch with its wonders. When you come across things that annoy or perplex you, curiosity is a far better response than hostility. Trying to understand the things you don't like can help you either to change them or to be more tolerant. *Being curious is like being mentally porous – you let things in; it's friendly.*

Generosity

Going back to Freud's idea of narcissism versus object love, or egotism versus altruism (see Chapter 3), it's important to be able to give and receive. Freud uses the idea of the amoeba that sends parts of itself out into the world and then sucks them back in again – perhaps along with some tasty bacteria. You can't send your whole self out – a core part of you has to remain or you wouldn't exist anymore – so you have to find a liveable balance between what you give and what you take, what you lose and what you keep. (And if Freud's not your favourite, there's always Nat King Cole's song "Nature Boy": "The greatest thing you'll ever learn/Is just to love and be loved in return.") Generosity follows on from curiosity in that it's a two-way street. Rather than focusing on what one can *get*, it might make life more interesting to think about what one can *give* – and one of the things you can give is to let other people be generous toward you.

It's an open secret that money doesn't make people happy. Still, people who have it (even the utterly miserable ones) are often very reluctant to give it up. But if enough of us truly grasped the idea that the point in life wasn't wealth acquisition – but maybe something more like dynamic exchange? – it would quickly make societies kinder, fairer, not to mention much more interesting. So rather than being overwhelmed by the prospect of having to dismantle a huge system, you can just try to dismantle the effects of that system in yourself.

The problem with imagining that material assets are the route to contentment is that it's hard to know when to stop – shouldn't you try to get a little bit more, just in case? This turns out to be a route to limitless dissatisfaction. The more you imagine you need, the more exploitable you become: you need money to buy more to make you feel safer and better, and so on. But everybody knows that's a rubbish way to live and the sooner we all get over it the better. With a very slight shift, it seems it would be possible to tip things the other way – toward generosity and away from selfishness (bearing in mind that you can't give *everything* away, and that your right to existence is as valid as anyone else's). Sometimes it can feel like society is suffering from a strange form of cognitive dissonance (i.e. holding mutually exclusive ideas and beliefs simultaneously) where all the biggest grossing, most fame-slavering movies and TV shows are all about the beauty of altruism – look at any war movie, disaster movie or romantic movie and you'll find representations of selflessness everywhere. There's huge consensus that immaterial rewards are better than material ones. So if we could just proceed from there, we'd have no trouble turning the climate crisis around. It would actually be really fun.

Social engagement

As we mentioned in Chapter 2, social engagement doesn't necessarily mean being a big, flamboyant show-off. Shy people are also excellent. Being socially engaged needn't mean persuading your entire village or borough to participate in a naked sit-in. You can be quietly effective, your small, kind acts radiating outwards. As George Eliot writes at the end of *Middlemarch*, "[T]he growing good of the world is partly dependent on unhistoric acts; and that things are not so ill with you and me as they might have been is half owing to the number who lived faithfully a hidden life, and rest in unvisited tombs."

Being socially engaged can mean any number of different things; the point is to be responsive and responsible. You might do this with the people in your immediate circle, and this might in turn help them to transmit thoughtfulness toward other people. Or you might get involved in some form of activism that's already up and running. You might think of something really helpful and necessary, and convince loads of other people to get behind it with you. You might give regular donations to organizations that help wildlife, or do a sponsored walk, or volunteer in a community garden. You might start being nicer to your ex. Anything with knock-on effects that can benefit others in big or small ways. By simply deciding to engage with the world on friendly terms, you make it a better place.

Ingenuity

Given that "business as usual" is killing the planet, we have to come up with unusual ideas if we want to save it. Of course, ideas like synthetic meat are interesting and might do a great deal for both animals and the environment, but not everyone's a biochemist.

In his essay "The Storyteller", the German cultural critic Walter Benjamin wrote that fairy tales can teach us to "meet the forces of the mythical world with cunning and high spirits".[25] The "mythical world", like the real world, is full of horrors, and stories have always helped us think about how to deal with that. The horror of environmental collapse is surely more terrifying than any troll or giant (although certain politicians do bring to mind caricature baddies), but Benjamin's advice is perhaps more pertinent than ever. We need to stay lively, and to think on our feet.

As with social engagement, you have to work within your means. If you're an economist or an architect, you might want to bring your special skills to the problem. That pretty much goes without saying. But if it seems that your skills have no immediate, large-scale application, you can still apply ecological ingenuity to any area of life. You can put up vintage curtains to seal off drafts in your house. You can think of brilliant things to do during the holidays that don't involve flying. You can make vegan food that's so delicious your friends will want to give up meat at once (good luck!). You can upcycle your clothes. Absolutely anything you can come up with. And, as with all the other qualities on the list, it will almost certainly also make life more satisfying and fun, reduce guilt and infect people around you. What could be more persuasive than witnessing other people actively enjoying being greener? Happy clappy? Yes. A viable plan? Also yes!

Honesty

This is another inward/outward trick – like breathing. Committing to being as honest as possible with yourself and other people (without wantonly hurting their feelings) is an ecological no-brainer. Climate scientists and journalists need to be honest with us, we need to be honest with ourselves about what we can do, and we need to be honest with other people about how we think and feel about climate breakdown. It's reached the point where it's no longer possible to pretend it isn't happening – even President Putin agrees that climate change is real; he just claims to believe it isn't anthropogenic, instead putting it down to "processes in the universe". Duh!

When you know where you stand, it's easier to make choices. If you're not quite sure whether or not your partner's cheating on you, you might not be able to decide what to do. Being told the truth, even if it's unwelcome, at least allows you to act according to your knowledge, and not be afraid you're just being rash or crazy. Having said that, you do need to be diplomatic about how you present climate change to children. There's no need to bludgeon them with alarming facts. They're going to find out about it somehow, so you need to be involved in that conversation. (There are more ideas about this in the following chapter.)

Radical friendliness

If you add up all of the things on this list, you end up with an outlook that could perhaps best be described as "radical friendliness" – a mode of thinking about, and interacting with, the world that can help stave off depression and anxiety by keeping you alert and engaged, and giving you a conceptual framework within which to act and make choices. It doesn't mean becoming ludicrously, falsely positive. Each "radically friendly" person would also be radically different. It's the very opposite of plastering on a fake smile while you're dying inside – the trick is to stay alive to yourself and others, to be as porous as you can bear to be without becoming waterlogged. At the same time as you don't deny suffering in yourself, you equally don't deny it in others. Their suffering is your suffering, and their happiness is your happiness. Of course, you will trip up, run into trouble and feel jealous, irritated or bored from time to time. This doesn't mean you're failing, it means you're noticing things. You'll always prefer some people to others, but that's just life. You can focus more on the people, and things, you like best without ostracizing everyone or everything else.

Climate-related suffering is terrible because of the guilt and rage (that inward/outward logic again). But fostering a forgiving, tolerant attitude toward yourself and others is absolutely essential to your future, and to the planet's. You can still speak truth to power, you can call out injustices, protest against wrongs, but always in a non-hateful way. Both XR and Greta Thunberg are major proponents of radical friendliness – they speak to the best parts of other people, even the ones they are criticizing. For instance, in her speech at the World Economic Forum's Annual Meeting in Davos in 2020, Greta Thunberg never uses the accusatory "you", but always the inclusive "we", as in: "We have to acknowledge that we have failed." In spite of her panicked intensity, she never descends into attack, but includes herself among the people who need to keep working to bring about constructive change. XR, too, insist on respect toward others, including those you massively disagree with, or who might not have your best interests at heart.

In a sense, radical friendliness aims to be the exact opposite of something like "McMindfulness" (a neoliberal bastardization of Buddhism). Rather than taking the non-ideological aspects of major religions, such as Buddhism, and retaining the selfish, uncritical bits ("meditation is good for you, therefore it's good for our business"), radical friendliness mainly interests itself with the more challenging, critical elements of so-called spiritual practices. While Trump-supporting Christian evangelists might want to build a wall against the Mexicans, anyone who actually gives a toss about the stories told in the New Testament (including the good old Christmas one) might clock that this is deeply un-Christian. Likewise, the Israeli government's treatment of Palestinians in occupied territories goes against the Old Testament's idea of extreme hospitality as told in the story of Abraham, who kept his house open on all sides. The Prophet Mohammed taught "Let the believer in God and the Day of Judgement honour his guest": a hospitable attitude toward others is a fundamental tenet of Islam. Basically, all major religions are full of injunctions to show kindness, generosity and respect toward other people. Of course, they do this due to human beings' tendencies to do the opposite – otherwise why bother to keep saying it? Still, while it's true to say that we're an incredibly troubled, volatile species, it's also true that we're brilliant at working together when we want to.

The fact that all cultures have developed stories and philosophies whose aim is to keep us on good terms with one another would suggest a certain consensus around the idea that this is desirable. Hard to sustain, maybe, but surely worth trying when the cost of *not* trying is so great. (And in case it sounds hard work and unrealistic, we'd only need to keep it up for the next decade while we deal with global heating, and then we can review the situation …) Basically, while religions get a bad rap for being morally restrictive and hypocritical, they also have the potential to be challenging and radical. It can sometimes seem as though the serious messages of religion and spirituality get lost behind the dull practice of routine, or co-opted for the purposes of social control. But the message is always in there somewhere: greed and self-interest don't lead anywhere worthwhile, while generosity and kindness do. Perhaps, at last, it might be time to start taking that seriously. In the words made famous by The Hollies, "He ain't heavy, he's my brother."

American Indian resilience

We will come back to this subject in more detail later, as it's been so important to people who are concerned with the psychological effects of climate change (see Chapter 9). I just wanted to introduce it briefly now as it relates to the question of resourcefulness, and to what sorts of things can *really* help when catastrophe strikes. If ecologically minded people are sometimes accused of being miserablists who want everyone to give up all pleasures ASAP, there is a very strong alternative current which suggests that developing a certain attitude, or set of character traits, is far more helpful and effective than stressing out over other people's coffee cups.

American Indians are in the shame-inducing position of being descended from people who lived in fully sustainable communities, but who had those communities destroyed by invaders who thought they knew better. These indigenous peoples are now being held up as masters of resilience by the very cultures that robbed them of their ways of life. They are suddenly being invited to teach us what they've learned from being the victims of genocide, in order to save us from being destroyed by an ecocide of our own making. But rather than getting bogged down in the excruciating injustice of it all, we urgently need to listen to what they've been trying all along to tell us. As the American Indian activist Nick Tilsen says, "If everybody lived like an American, it would take almost six planets to support life on earth. [...] We have to get back to one-planet living."[26] In support of this idea, he has been involved in developing a net-zero-energy community at Pine Ridge, Pennsylvania, consulting the community about how they would like to live and work. Pine Ridge inhabitants grow their own food, live in carbon-neutral spaces, and refrain from polluting their soil and water. There is no poverty in their community – they produce enough for everyone. As Tilsen tells it, they came up with the idea simply by asking themselves how they would like to live. By reclaiming their spiritualty and cultural identity, and allowing themselves to believe in the possibility of recovery, they dreamed, planned and built their way out of the extreme poverty and dissatisfaction of their reservation, in spite of repeatedly being told that what they were trying to do wasn't possible. In Tilsen's words again, "Our vision has to be at least as big as the challenges we're faced with."[27]

This kind of turning around of one's circumstances surely takes extreme generosity of spirit, as well as vision and determination. To put aside anger and resentment, and to find ways to transform a situation on one's own terms, is a sublime act of recovery. For the past few decades, "resilience" has been a catchphrase attached to American Indian communities, and hasn't always been welcome for obvious reasons. Who were you being resilient *for*? Are you helping your colonizers by being a "good Indian", thereby alleviating white guilt? Or can you recover for your own reasons, and on your own terms? This kind of question has been important for the kinds of psychological treatments offered to American Indians who were suffering from PTSD, depression or addictions. In Eduardo and Bonnie Duran's book *Native American Postcolonial Psychology*, they argue against coercive "cures" that basically try to tell people to pull their socks up and take responsibility, preferring more discursive, speculative spaces for exploring suffering. In particular, they seem to admire Jungian thinking, with its focus on symbolism, dreams and the unconscious.

While it's far from a direct analogy, it seems that younger generations, who haven't had a direct hand in creating the current environmental catastrophe, might be greatly helped by some of the key ideas that have come out of work around American Indian resilience. Namely, staying as physically healthy as possible, sustaining some kind of spiritual practice (which, for non-theists, might mean cultivating and maintaining good relationships), finding ways to regulate your mood, and having the idea that you can change things for the better – both for yourself and others. It would be totally understandable (indeed it *is* understandable) if the victims of extreme injustice descended into hate and rage against the architects of their misery, but it seems that many indigenous people, not to mention people born into unjust societies, or onto an already severely damaged planet, are sometimes able to find brilliant ways out of the impossible situations they've inherited. The trick is to keep believing it's possible, *and then to act on that belief.* And, in the light of this, we can now go on to discuss the fraught topic of children and climate change. Should you tell them about it? And, perhaps even more controversially, should you even be having them in the first place?

RESOURCES: CHAPTER OVERVIEW

1. You're not being paranoid: it really is true that there are people out there who don't want you to act on your environmental concerns. Recognizing this is helpful in that it removes the element of surprise when they do something to scupper you.
2. Get behind XR, even if you have doubts. They are the sort of people who will actually listen to you if you tell them about your concerns regarding their methods or their thinking.
3. Greta is an anagram of Great. Show support for her school strikes by joining them, or helping your kids (or other people's kids) to join in if they want to.
4. Cultivate your own brand of radical friendliness. The world could do with some good friends right now.
5. Don't obsess over the minutiae of carbon counts – yours or other people's. Try to develop a more open-ended, dynamic approach to responding to climate change.
6. Know that you will help others in the case of climate breakdown in whatever ways you can. Your experience through the lockdown of the COVID-19 crisis will already have shown you know how to do this.
7. Six fucking planets!!! It's not OK. Aim for Tilsen's "one-planet living" approach.
8. Resilience is for YOU. Never stay miserable in order to punish other people – in the end it will be you who suffers most.

6

BABIES, PARENTING AND CLIMATE CONVERSATIONS WITH CHILDREN

"Children are one third of our population and all of our future."

Select Panel for the Promotion of Child Health, 1981

Children have to be at the heart of any discussion about the environment, because they are the ones in the firing line. If you have them, work with them or hang out with them, there's the problem of what you can say to them about global heating. And if you don't have them, but think you might possibly want them, there's the question of whether it's responsible to bring new people into a world that may be on the verge of catastrophe.

In this chapter, we'll take a look at the thorny issue of being honest with children, at the same time as not overwhelming them with painful information. How can you find a middle ground, especially when you might be feeling panicked yourself? I'm afraid I'll be completely overriding the opinions of people who think that telling children about climate change is a form of child abuse, as I find this completely idiotic.[28] Children are going to hear about it one way or another, so you might as well give some thought to what you're going to tell them when they ask. While this seems like a difficult subject, it's probably slightly easier than the second topic – the ethics of procreation. And, of course, it's not just a question of whether the world is safe enough, but also whether it needs more humans in it. In my work, I often meet people in their twenties and thirties who are struggling with these questions head-on, unsure whether it's either safe or responsible to have a baby. On the bright side, there's much more of a public debate about it these days – conversations don't have to be limited to therapists' offices for fear of people accusing you of being melodramatic. Still, contemporary discussions on the subject demonstrate that there are no easy answers, which leaves people faced with an emotionally super-charged choice. At least if you have a vague grip on the various angles, you might feel more able to make your own decision based on a mixture of reason and emotion.

What Can You Say?

Sex and death are two notoriously difficult subjects for parents to discuss with their children. In the past, it may have been conventional to lie outright – to palm kids off with a load of dodgy stuff about storks and going to sleep in the sky. These days, we're more likely to try to find careful ways to let the truth filter through gradually – but children often jump ahead with questions based on things they've seen or heard. In the end, you can't control kids' knowledge, but you can be thoughtful about how to present them with painful truths. Like many children, I remember being very upset by the idea that the sun would destroy the earth at some unthinkable point in the future. How much worse must it be for today's kids to cope with the notion that we may reach irreversible ecological tipping points within the next decade?

Still, if children have no idea what's going on, how will they grasp the point of activities such as recycling and reducing waste? There are loads of well-made explanatory videos online, but of course there's something problematic about introducing your children to the natural world via the internet. However, if they have free access to computers they're likely to come across that stuff anyway; in which case, you might as well watch it too so you know what they're seeing and can answer questions about it. It's also helpful to see how other people have found ways to present potentially distressing content in a child-friendly manner, and to get some ideas on how to simplify the information without missing out the important bits.

From the very start, it's a good idea to try to give children direct experiences of natural environments, and to encourage them to enjoy and respect the natural world. Even most cities have wilder bits of parkland – you shouldn't need to go too far. Aquariums and city farms are also interesting for children, but they need to see plants and creatures in their native habitats. Once kids have the idea that there's a planet full of different lifeforms, you can begin to explain something about food chains, ecosystems and how the weather affects wildlife. Maybe some picture books (or even online videos) are helpful at this point. These sorts of connections can be quite mind-blowing, and a simple illustration or two might make the ideas easier to absorb.

By the time they've got the hang of all that, it's highly likely that kids will also have heard something about climate change – on the news, at school, while you were on the phone and thought they weren't listening. So now you can start to talk about needing to protect the environment from human activities, such as burning fuel and dumping waste. To keep things from becoming too distant and conceptual, it can help to focus on a particular type of animal that the child likes – even if it's one they've never seen in real life, for example, penguins or lions. You can explain how penguins need sea ice, or that droughts and floods can make lions sick. Of course, you'd only do this once a child already has lots of questions of their own about climate change. The idea isn't to traumatize them by telling them that their favourite animal is going to die; it's to simplify the information so they get a clearer picture of how different things are linked, such as exhaust fumes and icebergs. Zoos and aquariums often use this method in their teaching programmes, as it helps to make things seem tangible.

At some point, it's almost inevitable that children will be unsettled, if not outright horrified. They might ask questions that will test your ability to present the information kindly and calmly. A friend described talking to his six-year-old about climate change. The conversation quickly turned into a series of questions along the lines of "Will lots of animals die?" to "Will lots of people die?" to "Will *we* die?" accompanied by a rising sense of panic on both sides. It's important to stay as calm as possible when discussing climate change with kids, and to emphasize the fact that there are things people can do about it – and that *you* are very much one of those people. Definitely don't give children the idea that it will be up to their generation to solve the issue. Although this is almost certainly going to turn out to be the case, they don't need to know it when they're three. It's more helpful to tell them about all the things people are already doing to help improve the situation. They needn't feel the enormous burden of responsibility right from the start. The idea should be more that all sorts of people are already working on it, and that children can join in and do their bit too.

If you drip-feed the information, it's likely to go much better than if you lay it all on the line at once. Like dealing with the subjects of sex and death, it's vital to be responsive and to wait until kids have their own questions or concerns, rather than to steam in according to your timetable. Children need to know they can check in with you with their questions and worries whenever they feel like it, and that you will answer them in a friendly and truthful way. If there are things you don't know, you can always look them up – maybe even together. It's comforting for children to feel that it's a subject which the adults around them care about and take seriously. Of course, children are clever and however brilliant you are at appearing cool, calm and collected, they will inevitably pick up on your anxieties. This isn't because you have failed, but because they are attuned to you. They know perfectly well that you may pretend to be happy when you're sad, or placid when you're angry. So don't be upset if your kids clock you as a worrier – that's just part of being a real parent, rather than one in an advert. Managing your children's climate fears is like any other aspect of parenting – you're bound to fuck it up in places. As the therapist Donald Winnicott explained in the 1950s, the point isn't to be perfect, merely "good enough". Parents who believe in perfection, either for themselves or their children, are *awful.*

Over-zealous adults who become freaked out themselves can easily pass on their anxieties to children by being overtly hysterical about anything from food waste to recycling. Because children don't – and can't – have a nuanced view of the big picture, they might feel like any little scrap of paper they fail to deliver to the correct bin is going to cause a catastrophe. If the adult in question is their parent or teacher, they may become so anxious to please that person that they will go overboard, to the point of becoming obsessional.

If a child in your care starts to show obsessional symptoms, like rituals, checking or extreme punctiliousness around environmental activities, try not to panic. Obsessional symptoms are generally attempts at self-cure; the child may believe they can prevent bad things from happening using semi-magical means – "The world will be OK so long as I recycle every piece of plastic I find in the street." You can try to reassure them that *they are not solely responsible for the whole planet.* These days, I think you can also reasonably say that climate denial is becoming much rarer. Up until recently, deniers have been given far too much of a platform, but this has become less and less tenable. So you won't be altogether lying if you give children hope that things will get better. (See Chapter 10 for concrete signs of improvement. Or read Lily Cole's book *Who Cares Wins: Reasons For Optimism In Our Changing World.*) As well as being given the facts, children need to be firmly reassured that all is not lost and that there's still everything to play for.

Lastly, perhaps, it's helpful to make environmental activities as fun and stress-free as possible. *Never* shout at your child for wasting food/water/heat. Nor for wanting crappy plastic toys, or for wanting to eat at McDonald's. And definitely not for wanting to fit in with friends whose parents might not be as ecologically minded as you are. You can talk to children carefully and gently about all of this stuff, but the last thing you want is to set them against you, or to force them to keep secrets. It makes more sense to play a long game, sometimes letting your child do things you disapprove of, and to allow them to make their own choices. You have to trust that the arguments in favour of looking after the environment are so good that you don't need to force them on anyone, especially not small kids. What you generally find is that children are more than willing environmentalists. They totally and intuitively see the point in things like not eating animals. In fact, it probably takes a great deal of subterfuge and brainwashing to desensitize children to "normal" activities like meat-eating. They might slip up, or want things that go against ecological reason, but don't we all? If you see it as your job to help them find the best possible way through a difficult problem, it's highly likely that they'll end up being much better at the whole business than you are. They're growing up in a world where veganism is considered mainstream, and flight-shaming is actually a thing. So if you just let kids get on with it without too much fuss, the chances are you'll be pleasantly surprised (if not unpleasantly embarrassed when they begin to pick you up on things you'd never even thought of).

The best thing you can do is to teach by example, and to show children that it's no sweat to be nice to the planet. Whether you do special, sit-down activities, like making a shopping bag out of strips of old clothes, or just enjoy everyday activities like putting the washing on a line or drying rack instead of in a tumble drier, they just need to know that you're on the case, and they can be too.

To Breed or Not to Breed?

This question is the main reason I wanted to write this book, but I now realize how naive that was. It's a subject so loaded with pain that it feels almost unbearable to think about. It's come up again and again in my consulting room over the last five or so years: "Is it wrong to have a baby?" I have a nineteen-year-old daughter and it seems to me that she and her friends are in pretty unanimous agreement that the answer to this question is "yes". Perhaps Gen Zers are in the lucky position, so far, of not being faced with the real difficulty of the choice. They're not yet in relationships and life situations where the question has become pressing and personal. But for millennials it's very real, and often hugely upsetting. And for the *parents* of Gen Zers and millennials, it's extremely distressing too. Of course you can't, and shouldn't, bank on your child making you a grandparent, but it's deeply saddening to see the option being batted out of the park before there's even been time to look at it.

The fact that the question comes in two mutually reinforcing parts makes it absolutely devastating. There's the environmental impact of the child on the world, and the potentially horrifying impact of climate breakdown on the child. Put the two together and it's easy to see why so many people are deciding that it's unjustifiable to breed.

Embedded in the decision over whether or not to have a baby are further questions such as: "Is it reasonable to hope?" and "Is life worth living? And, if so, what's it for?" Even, "If it turned out that I was going to be one of the last humans on earth, would I wish I hadn't been born?" While it might appear to many to make logical sense not to reproduce, that doesn't mean the idea is free from profound emotional impact. Can we coolly concede that the world is no longer a viable human habitat, and get on with living out the rest of our lives as harmlessly as possible? I would say not, or at least not for many people. If you like the idea of having children, the prospect of abstaining is just too sad.

For Meghan Kallman and Josephine Ferorelli, the founders of Conceivable Future – an American, female-led collective – the fact that large numbers of people are asking these questions *at all* is already proof of the urgency of the situation. In a sense, it doesn't matter which side you come down on, the tragedy is already there. So, Kallman and Ferorelli argue, while individuals go to work on making these decisions for themselves, those in a position to legislate and make big, sweeping changes need to be acting in order to de-escalate the situation. For instance, governments ought to be making rail travel cheaper and flying more expensive – it shouldn't be down to young couples to solve the climate crisis by forgoing one of life's most potentially enriching, satisfying experiences. The ultimate point of an organization such as Conceivable Futures isn't to advise people on whether or not to have a baby, it's to raise awareness around the subject so that governments feel more pressure to act to protect the planet.

BirthStrike is another growing movement, whose approach could be said to be more strident: supporters won't be having any babies until governments can demonstrate that they are taking the crisis seriously. Founded by Blythe Pepino, a British musician, the movement and its members have experienced a terrible backlash, which just goes to show how much some people think that women ought to be breeding, or else. Pepino herself has been trolled by virulent haters, using all the usual misogynist threats and insults to bully her into shutting up. Without wanting to excuse this sort of behaviour for a millisecond, I suppose you can at least say it demonstrates what a super-charged subject this is.

The choice not to have children doesn't only impact on the feelings of a bunch of affluent thirty-somethings. Reproduction is an important factor in the shape and direction of an economy. In spite of relative economic stability, birth rates in many developed countries are declining (although America saw a tiny 0.09 percent increase in 2019, after four years of falling figures). The reasons for this are complex and varied – it certainly can't solely be put down to nervousness around climate change. In many developed countries, it's becoming increasingly hard for younger people to set themselves up in careers and stable housing. There may also be other factors, like cultural shifts away from ideas around tradition and continuity, and toward individual happiness. If you're going to have a child you're probably going to have to really want it. Getting pregnant by mistake is less and less of a thing.

Perhaps counter-intuitively, wealth and low birth rates often go together. In countries with a high standard of living and good medicine, people might have fewer children but expect more for them. They will want them to have the finest educations and to aim for the bougiest forms of work. Life becomes about connoisseurship rather than survival – a luxury experience rather than a hard grind. One lives for oneself rather than for the family or community. It's precisely these sorts of societies that are causing problems for the planet. It's not poor people in Africa who are supposedly overpopulating and spilling out all over the place. It's careful, rich people with 1.87 children (or those who are so prosperous and comfortable they have 3 or more) who constantly upgrade their iPhones, have nice freezers and go on yoga retreats in the Himalayas. So, when people say that the best thing you can do for the planet is to have one less baby, it matters very much to whom this baby is born. An Australian Aboriginal baby has quite a different carbon footprint from, say, a middle-class British one.

The otherwise largely saintly David Attenborough has got himself into trouble over the question of overpopulation by suggesting that it's our urgent responsibility as individuals to breed less. One problem with this is that it touches on the extremely fraught topic of race and women's reproductive rights. Attenborough himself apparently couches his argument within feminism. (A privileged, non-intersectional version of it.) He tells us: "Wherever women are given political control of their bodies, where they have the vote, education, appropriate medical facilities and they can read and have rights and so on, the birth rate falls – there's no exceptions to that."[29] And, for Attenborough, this is simply good. Fewer people equals less pollution. It's straightforward, numerical and surely well-meaning. Still, the first problem with it is that it lands the problem of saving the planet squarely on the shoulders of women in developing countries, as they are the ones who may currently lack access to education and birth control. So educated women in wealthy, super-consumerist societies can guiltlessly carry on as we are, even though each one of our babies will consume many times more than any of those superfluous African babies that Attenborough is apparently trying to save us from.

The second problem is less immediately apparent, but intimately connected to the first. It has to do with the effects of falling populations on societies. If things were as straightforward as David Attenborough seems to suggest, women would be given rights and contraceptives, numbers of humans would quickly decrease, less stuff would get used, and our climate and ecosystems could begin to recover. As it turns out, declining populations don't just keep going as they are, but on a smaller, more planet-friendly scale. Instead, shifting demographics can cause major social upheavals. As historian Trent MacNamara argues in his fascinating article "Liberal Societies Have Dangerously Low Birth Rates": "Low birth rates [...] threaten welfare states with bankruptcy, and nations with the destabilizing politics of cultural extinction."[30] Why would this be? In ageing populations – where people live for longer, and aren't replaced by younger generations – there will be fewer people of working age supporting more people of retirement age. As this becomes unsustainable, it becomes necessary to invite immigrants in. As it tends to be wealthier countries that find themselves in this predicament, it may not be hard to persuade people living in poorer countries to migrate, with the promise of a more stable, swanky lifestyle. The new arrivals might even things up for a bit, but migrants tend quickly to adopt the reproductive habits of the surrounding population. Within a couple of generations, everyone's back to where they were. This is because affluence may very well depend on smaller family sizes, which in turn may be accompanied by high childcare costs and expensive housing. The new people want what the old people have and, in order to get it, they have to make the same sacrifices. So a next wave of migrants needs to be sought out to come and fill in the gaps at entry level. This might sound great, in theory – the first generation of immigrants are happily settled, making way for a new lot of people to be rescued from poverty. Meanwhile, birth rates are diminishing, leading to lower consumption, and everyone has access to an admirable cultural life. Yay! The problem is then that the original population may, at this point, begin to worry that they are committing so-called "race suicide". They've been good, amassed money and built a wonderful society that everyone supposedly wants to be a part of, but now they fear becoming a minority within it. From here you potentially get disturbing developments like ethno-nationalist movements, "replacement theory" (the idea that the white race is gradually being usurped) and other forms of far-right, racist claptrap.

As ever, we're talking about complex systems and speculative theories, but the basic point is that declining to have babies isn't necessarily a straightforwardly good thing if what you end up with is a super-consumerist society run by white nationalists. Instead of this, it might be better to have a baby, if that's what you want, and to bring it up to be an anti-racist environmentalist.

The problem with babies isn't the babies themselves, but the things people in developed countries associate with them. For instance, having a family needn't mean owning a big "family" car. Or even any car – especially not if you live in a large town or city. It also needn't mean living in an enormous, over-heated house, using a ton of non-biodegradable disposable nappies, dressing your kids in a constant parade of new clothes, and eating meat. The problem with the extra person isn't the person, but the stuff extraneous to that person. Unfortunately, neo-capitalist societies are liable to instil in you the sense that if you don't provide your child with all that crap, then you're not exactly a great parent. Love is all very well, but you're also supposed to express it through the luxuries you provide, otherwise what's it really worth?

To go back to Trent MacNamara's lovely *Atlantic* article, he concludes by asking what might be the ultimate outcome of the possibility that "free and egalitarian societies" aren't very good at reproducing themselves. They're great for a bit, but the people run out, the resources pile up unevenly, and nobody can quite see the point in trying to keep it all going. Most miserably, then, you get things like the Voluntary Human Extinction Movement – people for whom human life is so debased and pointless that it would be better if it just stopped happening altogether. Thankfully, MacNamara's take is far more optimistic. While it might be the case that affluent, industrialized societies can start to present us with dead ends, rather than giving up this could make it "[m]ore likely [that] some new idea will arise among us to dignify and eternalize our lives and way of life. That new world could retain the best of our blessings."

What this new world might be is anyone's guess, but we shouldn't be put off by the idea that it can't start until someone comes up with a brilliantly subtle economic theory that will painlessly supersede consumer capitalism. Of course, somebody might do just that – in fact, who's to say that the UK-based left-wing media organization Novara Media haven't already done it with "fully automated luxury communism"? Or perhaps coronavirus has already triggered the change that political activists and theorists have failed to ignite? Still, until whatever it is kicks in, we can surely get a head start with simple ideas along the lines of: "Less stuff, more love. Keep going." You don't need a PhD from the London School of Economics to see that it might actually make a difference.

CHILDREN AND BABIES: CHAPTER OVERVIEW

1. Be aware of your anxieties and try not to pass your eco-anxiety in its complete adult form on to children. Contain it for their sakes.
2. Still, there's also room for careful honesty – children hate being lied to.
3. Teach environmentalism by example, and make it look as enjoyable and effortless as you can.
4. Be there to talk to kids about the climate crisis when they need to – but don't worry if you don't have all the answers. It's not one conversation, it's a lifelong project.
5. Remember to focus on the good news.
6. If you really want to have a baby, don't be put off. The arguments for and against aren't simple, so you might as well take a risk and go with your feelings.
7. If you know you don't want to have a baby, lucky you! That will save you a whole lot of agonizing. Still, *never* be judgemental about other people's decision to go ahead with it. The Human Extinction Movement is bullshit, and you might end up with fascism instead.
8. Babies need love, not designer lifestyles. It's consumerism, not babies, that we need to place limits on. David Attenborough knows this – he can still be the nation's daddy.

7
PLEASURE YOURSELF: YOU KNOW YOU WANT TO

"The greatest sweetener of human life is Friendship. To raise this to the highest pitch of enjoyment, is a secret which but few discover."

Joseph Addison, writer and politician

It's absolutely vital that environmentalism doesn't equate with miserabilism. The point isn't to accept that things have to be sad and dull, and to somehow get off on it. It's true that ecology invites people to give up a number of enjoyable habits for the sake of the greater good, but this needn't mean becoming a sour-faced disapproval-monger.

In this chapter, we'll have a look at some of the potentially polluting activities that lots of us modern humans have been encouraged to expect and take for granted as a normal part of an enjoyable life – fashion, fancy food, holidays, driving, looking at screens. Clearly, a lot of these normal pleasures have been taken away in an instant by the coronavirus pandemic, and as yet it's hard to know in what form they will be re-established into our lives. It may feel a bit like watching a movie and observing with curiosity our "old" social ways of kissing and in-person camaraderie, this chapter looks to how we can still enjoy these things, if they still exist beyond May 2020, but in a more environmentally responsible way. It won't just be a case of "do clothes swaps, go vegan, stay in your neighbourhood, ditch your car and learn some card games".

These are all good things to do, but you don't necessarily want to do all of them at all times. In fact, trying to be too stringent can make you fall off the wagon altogether.

We'll also have a look at the guilt that can come from "slip-ups", and how perfectionism can cause anxiety – which can, in turn, make you feel like giving up. It's easy to feel like a fake or a hypocrite if you go rogue once in a while, but it's better to try and fail than not to try at all. To err is human, after all.

Finally, we'll tackle some of the criticisms you might receive from people who seem to want to undermine your efforts, along the lines of: "Don't you know that avocados are worse for the planet than organic, locally-produced meat?" and "Synthetic shoes are even less eco-friendly than leather ones." These may be valid points, but if people's secret agenda is to deflate you in order to make themselves feel less guilty about their own habits, maybe you need to be able to defend yourself (even just internally). So, although this chapter will focus on pleasure, it will also inevitably glance at guilt, because the two so often go together.

Clothes

I'll get this one out of the way first as it's my Achilles' heel. Because it's the area I'm most personally bothered about, it will contain prototypes of arguments that I believe are also applicable to other topics. I know that there are people who like meat and fish as much as I like fashion, so I hope that some of the stuff here will be translatable.

I properly, pathologically LOVE clothes. Worse than that, I love the way they change over time, and that things go in and out of fashion. This is the worst possible way to like clothes, because it means you tend to want new ones before you've worn out the old ones. It also means you do dumb things like buy *Vogue*, watch catwalk shows on YouTube, and follow @ diet_prada on Instagram. Not owning a car is no sacrifice to me, but it feels impossible to stop enjoying clothes.

But fashion is no joke. The clothing industry produces 10 percent of the world's carbon emissions – *four times more than flying*. And this is perhaps a conservative estimate, as it may not take into account all aspects of the industry, such as the distribution of garments as well as their production. Fashion pollutes through the manufacture of fabrics, the transportation of garments, and the production of waste. Growing one kilo of cotton requires 10,000 litres of water, not to mention a good dousing of pesticides. The use of viscose promotes deforestation. Polyester is basically a fossil fuel. Wool is intimately linked with the meat industry, and its production may involve cruel practices such as "mulesing" – skinning the live sheep's backside (without anaesthesia) to discourage flystrike. These fabrics may then be dyed, causing damage to rivers, after which they will be made into garments (perhaps by workers with poorly enforced labour rights) and often transported long-distance before being worn once or twice and eventually ending up as landfill. Tights are apparently particularly terrible as they are mostly non-biodegradable and very easily damaged, so they produce tons of the worst kind of waste.

And fashion is bad in other ways. It reinforces unrealistic and persecutory beauty ideals. Although, to be fair, it also finds beauty in things that are often considered "ugly", and it can be subversive, challenging and weirdly democratic. For instance, a broke "ugly" person may be deemed cooler than a rich "beautiful" one. Or "ordinariness" might suddenly be valued over "extraordinariness". It's a very strange, unpredictable system, and perhaps that's what makes it so addictive, like gambling.

However, there are obviously more and less heinous ways to enjoy dressing up, so this will be an informal guide to the least worst ways to be a fashion hag.

Firstly, avoid the high street as much as possible. See my opening comments, it's hard to know if the high street will still exist at the time of reading. Buying fast, mass-produced fashion is pretty much completely bad. I know some high street shops have "ethical" lines, but that just goes to show how unethical the rest of their stuff is. It's good that they're trying, but they need to try quite a lot harder. The same goes for online fast-fashion. Of course, online or high street stuff might be clean and respectable-looking for work, not to mention affordable and accessible, so you may feel you have no choice. If you have to buy clothes on the high street, you can try to keep it to a minimum and be nice to the things you buy so that they last longer. One trick I only learned recently was always to use an apron for cooking so you don't mess your clothes up with oil and tomato stains. Even more drastic is to change into a scruffy outfit immediately when you get home to make it less likely that you'll damage your clothes while cooking, eating or cleaning. *Home* is often much more pernicious to clothes than *out*. Also, washing things in cooler water and air drying (as opposed to tumble drying) makes fabrics far less likely to shrink and bobble. And wearing sockettes under your tights (to protect the toe area from holes) can make them last much longer. If you buy thicker tights, so they don't snag so easily, and do this trick, you can make them last for years! Also, cut your toenails. And when your tights finally do give out, you can give them a new lease of life by snipping off the toes and turning them into leggings. If you do all this, you might even be able to justify the odd expensive pair. If you need new stuff, you can look out for ethical brands like ThreadUp, Alternative Apparel and People Tree, who recycle, use organic fibres and pay their workers properly.

Until things change, however, second-hand shopping seems to me by far the most appealing way to get hold of exciting clothes while causing minimal damage. Having said that, even within the second-hand market, there are gradations of goodness. Obviously charity shops are great because the benefits are manifold: you get cheap clothes, other people get to Marie Kondo their houses, the charity gets your cash, and the clothes are saved from landfill. It does take time to find things, though, and the stuff you like may not be in your size. Charity shopping is amazing for people like me who happen to love it – and who can sew – but may not be to everyone's taste.

Then there are sites such as eBay, which solve a few of the problems of charity shopping, in that you can search for particular garments in your size. Also, you can search for things from all over the world, although this adds air miles to your purchases. This huge variety is great because, if you go and see, say, the movie *Little Women*, and realize that the future of fashion is waistcoats, you can immediately go online and find the perfect Jo March gilet in your size and colour. You don't even need to wait a few months for the high street to catch on. For people who are obsessed with the flightiness of fashion, but who try to be ecologically decent, eBay is the best invention ever. It has its limits – it relies on fuel-consuming delivery systems for a start, but we'll get on to the comparison between real-world shops and online shopping shortly. You could also argue that things like eBay, Depop and even charity shops encourage people to buy more because they know that there's something useful they can do with the clothes afterwards. I try to curb thoughts like this as they can lead to horrible, spiralling obsessionality. Still, I suppose the trick is to remember that too much of a good thing can stop it being a good thing. Even eBay has its limits, so try not to over-shop.

Another option is designer second-hand. This is sold through selective websites such as Vestiare Collective, in real-world designer exchange shops and, for the moment at least, in some department stores (although it will be interesting to see whether this catches on). This can be great if you're looking for something really special, or if you just have expensive taste. The downside of it is that it's a great guilt-reliever, but actually probably just encourages the kinds of toxic shopping behaviours that make fashion a major polluter. While the clothes being sold in the "goody-goody" section of a department store may be referred to as "pre-loved", the last thing they want are garments that show any signs of having been loved or even liked. Often the rules for sellers will specify "shoes that have only been worn once", for instance. Only articles from designers' main lines are accepted – no diffusion ranges – and the tiniest imperfection, even on the lining, will disqualify an item from being accepted. So, basically, only oligarch's daughters who buy more items from Gucci than they have parties to go to need apply.

Perhaps a more egalitarian, communal means of accessing high-end stuff are the rental services that are gradually springing up, where you can get a gown for an evening, or even a fortnight, and then give it back. Given that Instagram has made "rewearing" an issue – who knew we needed a special verb to describe wearing our clothes more than once? – rentals are a brilliant way to keep things lively without bingeing. And it's not just party dresses for hire: there are also companies that rent out things like Balenciaga puffers and limited edition streetwear.

Also, clothes swaps are good, in theory, for all the obvious reasons. Still, I've never done it because I have such a sentimental attachment to my clothes that I don't think I could watch while other people took them away from me. I either have to give them to a very deserving charity (such as Traid) and then run away as fast as possible after the drop-off, or sell them in dress agencies where I know the shop-owners well enough to ask for things back if I have sudden regrets.

Basically, fashion is a huge and dreadful system, and intersecting with it at any point is problematic. In my household it goes like this: my daughter looks at cool Instagram accounts that I've never heard of, and then finds clothes in charity shops that approximate what she sees. Designers observe what the young folk are wearing and copy it, quickly followed by the high street. I admire what the designers make, but can't afford it. I wait a few years for the oligarch's daughters to sell off their unworn stuff in snotty dress agencies, and then for the second owners to sell it on eBay. At this point I buy it, by which time it's old enough to have become interesting to Korean Instagrammers, so my daughter steals it from me. What I'm trying to say is that I see there's no perfect way to like fashion – the whole thing is basically terrible and everyone who's into it is culpable, including post-apocalyptic dressers like me. But maybe if there's something you really can't give up, you kind of have to let yourself have a bit of it. If you're vegetarian/vegan, grow your own vegetables, don't own a car and are stingy with your central heating, perhaps you are allowed the odd vintage Margiela purchase? (Translate self-justifying logic to meat-eating/car-owning as appropriate.) It's all about offsetting.

Of course, these days, designers and large manufacturers are catching on to the fact that everyone's worried about the planet, especially young people. (Well, not quite everyone…) So there are more and more options to buy vegan trainers with soles made from recycled bottles, and T-shirts made from organic cotton, and so on. It's a really good start, but *buying less* is the most important thing of all. So when Adidas launches a range of ethical trainers, as well as being delighted, you might also remember that they are basically doing this because they don't want you to stop shopping.

Also, while trying to make worthwhile environmental choices, it's interesting to see how different designers and makers come at the same problem. Two of my most frequent eBay searches are for pieces designed by Vivienne Westwood and Issey Miyake, both of whom put a great deal of thought into responsible production. (I'd buy their stuff first-hand if I could afford it.) Westwood uses carefully sourced materials, natural dyes and fair-trade manufacture, and an approach that discourages over-shopping. Her company constantly reworks classic designs so that seasons blend and match with each other, rather than being in this year, out the next. However, some of her lovely natural fabrics require dry cleaning and are susceptible to moths, plus the more fitted pieces require you to stay the same size. On the other hand, and quite counter-intuitively, Issey Miyake works with indestructible man-made fabrics that require very little maintenance – not even ironing or steaming. One piece will last not only your lifetime, but also the lifetimes of any lucky friends or relatives who stand to inherit your wardrobe. No need to send things to landfill because the clothes are so beautiful and easy to wear (they even reduce and expand to accommodate your changing body) that they will always find a grateful owner. But please don't actually start to look for these clothes on eBay as then we will be bidding against one another.

Other designers are also beginning to rethink their production methods; for instance, Maison Margiela has mixed upcycled pieces sourced from thrift shops with new designs, while Marni and Pringle have made collections from deadstock (old fabrics and garments that didn't sell). And, of course, Stella McCartney has been on to all this from day one. There are also newer designers working with novel ideas such as coating clothes with carbon-absorbing algae, but this does smack a little of greenwashing – why not put the algae elsewhere and stop producing so many new clothes? Anyhow, fashion people are definitely on the case, trying to work out how to stay in business in a world that increasingly disapproves of pollution and over-consumption, and it will undoubtedly be interesting to see what else they come up with.

All of which is to say that buying boring stuff from companies whose only claim to fame is their eco-credentials is far from the only way to go.

Don't Get Mad, Get Eco

Other fashion conundrums involve questions around cruelty-free synthetics versus biodegradable leather, or new fake fur versus real vintage. I was once spat on and shouted at for wearing a not particularly realistic biodegradable fake fur. When I explained this to the person shouting at me, thinking I had all bases covered, they said, "So what, you're encouraging other people to wear the real thing." This is the sort of hypercritical, paranoiac thinking that I'm very afraid of entering into as it makes life miserable for everyone. And it's not hard to see how easy it would be to fall into. In my experience, the biggest obstacle to my own environmental engagement has been self-punishing internal voices that tell me I'm a big phoney and should probably just give up. One option would be to listen to these voices and pull my socks up, but the problem with this is that they don't stop, they just find something new to latch onto. I think an over-attention to secret self-flagellation is also what leads people to feel licensed to spit on other people in public places. If you're stringently living according to the laws of your own inner persecutor, why should other people get away with being slack? This outlook is totally antithetical to radical friendliness and, though it may frighten the odd person out of ever wearing their favourite coat again, probably does very little to bring people on board, or to make environmentalism and animal rights look good in any way.

Far better to be flexible, open-minded and see what you can do. I only buy shoes second-hand, so allow myself either leather or synthetic to balance out the damage. When all the second-hand shoes in the world have run out, I will be delighted to buy new ones made from pineapple leather (although increased pineapple farming may be wreaking ecological havoc by then). This is at least the sort of imperfect solution that helps me to stay sane and therefore seems far more serviceable than becoming a sanctimonious eco-bot.

At the other end of this spectrum are the people who don't seem to give a fuck about taking care of the planet, and who have a go at you for your fallible efforts. The anti-avocado meat-eaters. Often these people are otherwise delightful – they might even be your friends, family or lovers. Perhaps the best way to tackle them is sweetly. They are just trying to make themselves feel better by demonstrating that these problems are far too complex and entrenched for single solutions to work. They are simply pointing out that if all of us jump on the same solutions at the same time – swapping cows' milk for almond milk, say – we just end up creating new crises. (Intensive almond farming is killing bees, and bees are key to life on earth. I know you know this, so I'll refrain from ecosplaining.) The answer to this clearly isn't to give up, but to be open to diverse solutions, and to a certain amount of flip-flopping. Just know, in your heart, that being imperfect – and friendly – is likely to be far, far better for you, for other people, and for bees, than any rigid, punitive orthodoxy. I hereby license you to slip up in your own heterogeneous ways. Have fun!

Shops or Clicks?

This one's definitely applicable to food as well as to clothes, not to mention toys and maybe even technology. Is it better to do your shopping online, or to physically go to real-world shops? Apparently the answer is very unclear. Studies have to take so many variables into account that it turns out to be difficult to come down definitively on either side. Both online shops and real ones have single or multiple locations where goods are dropped, so up to this point there's no difference (except between individual retailers, who may be more or less environmentally aware). From here, the questions begin to focus on individual shoppers' behaviours, which are in turn affected by their locations and modes of transport. If you live in the countryside, or in a suburb, a trip to the shops may very well involve driving. Still, an online purchase will very often involve someone driving to you. However, courier companies use software to minimize driving distances and save fuel (a nice instance where profits and eco-benefits dovetail one another). Individual shoppers may not be quite so canny – although surely most people do actually *try* to make their shopping trips as hassle-free as possible by doing things in a sensible order. Then there's the question of what you do with the time you've saved by ordering online. American studies seemed to show that people who did their shopping online might then very well use that extra hour to drive over to a friend's house. If this all sounds a bit imprecise and speculative, they worked this out by measuring traffic volumes in places, and at times, of increased online shopping and discovered that high instances of online shopping made no predictable impact on the numbers of cars on roads.[31] Sometimes high clicking levels equated with high driving levels.

There are also questions about the building, heating, decorating and cleaning of fancy retail spaces over functional warehouses, and details like this lead people to lean slightly toward the idea that online shopping may win by a narrow margin. But then there are also differences of location; if you live in a town or city with good public transport, or if you ride a bicycle, or even drive a mega-green car, then the balance starts to tip toward real-world shops. Then there are other factors, like returns. If you shop in the real world you may be less likely to buy speculatively, thinking you can always give it back later. For online companies, returns have become a real issue – especially with the added peril of people simply ordering clothes for their Instagram selfies with no intention whatsoever of keeping them. Needless to say, causing companies to drive, or fly, clothes around for no good reason is a problematic way to conduct yourself. Just because you don't actually buy the clothes, it doesn't follow that you are being virtuous and abstemious.

Aside from all this, while we're talking about orthodoxies and complex systems, and the fact that for all actions there is an equal and opposite reaction, shops have so far been a really important part of human cultures. They are social spaces and, as such, militate against isolation and atomization. Trade is where different cultures intersect, which can promote tolerance and diversity. Going out there to get stuff, or putting yourself out there to sell stuff, can be a massively pleasurable way to engage. People need one another, and shopping is a place where this becomes abundantly obvious. Losing real-world shops is like losing breeds of insect – you might not see how important they were till after they're gone. So while the environmental message might be to consume less, shopping surely still has a place as a vital social activity. As Vivienne Westwood likes to remind us, "Buy less, choose well, make it last."[32]

Whining and dining

This is a place where all the problems outlined in the section above can really kick into action – perhaps even more so, because food is a primal human need. Many of us could probably live out the rest of our lives in the clothes we already own, but food is an everyday necessity. More than that, it's always bound to be emotionally loaded – our very first relationships are conducted around food (a.k.a. milk), after all. If you give up, or limit, your satisfactions in this area, it can sometimes have a surprising impact on your emotions or even on your mental health. For instance, "orthorexia" – an eating disorder that involves an obsessive insistence on eating "correctly" – can, perhaps surprisingly, make people very unhealthy. Especially if you consider anxiety and misery to be legitimate forms of "unwellness", before you even get onto things such as vitamin deficiency or excessive weight loss. Maintaining a good relationship with food, like getting enough sleep, is important to being a functional environmentalist. If being a "good" person seems to involve becoming depressed or malnourished, it's perhaps time to ask yourself what "goodness" means to you. If you wouldn't impose it on other people, definitely don't impose it on yourself.

Being vegan is thought to be one of the best – if not *the* best – thing you can do for the planet. However, it seems that not all of us are able to sustain it for a lifetime. I've done spells of it on and off, sometimes for years, and try to be mostly vegan at home, but I'm quite a way off 100 percent. I feel terrible for saying it, because it seems weak, but I've also had mild eating disorders, plus a tendency to fret and obsess, so I've decided I'm best off being midway between vegetarian and vegan for now.

As with clothes, it seems to me that a bit of leniency can be essential to keeping you on the straight and narrow. It's true that wavering, like I do, can lead to a certain amount of cognitive dissonance – I really do understand that the meat and dairy industries are inextricable – but I also believe that aiming for complete mental, physical and behavioural hygiene is a losing game.

As I understand it, there are very few people who would begrudge an Inuit their right to fish. Still, you can't simply opt to be an Inuit if that's not the hand you were dealt. So, as with everything, we all need to cut each other – and ourselves – a bit of slack. It may turn out that there is room on this earth for both small, traditional indigenous communities and high-tech metropolises. I really hope so. And, if there is, it will probably be thanks to a certain amount of fellow feeling flowing in numerous directions. If you LOVE steak, and can handle the idea of a cow being killed and rainforests cut down to clear land for the sake of your enjoyment, maybe it'll be better for you in the long run to allow yourself to eat it (consciously choosing cows who have at least lived more locally to you), even just once in a while, at least until the synthetic stuff goes mainstream. And maybe if your partner or friends want to do this, but you don't, it's wise to let them go ahead without a fuss.

There's a free-floating cultural myth that "proper" vegans and vegetarians are total purists who wouldn't let a particle of contraband foodstuff breach the boundaries of their bodies. This characterization comes through in unhappy instances such as the news story of the angry pizza chef who, in 2017, "spiked" a vegan customer's dinner before posting on Facebook: "Pious, judgemental vegan (who I spent all day cooking for) has gone to bed, still believing she's a vegan."[33] Apart from all the obvious problems with this, there's the fact that this chef seemed to think that veganism was more akin to a form of religious mysticism than a practical, rational choice. Not all vegans are interested in their own physical and metaphysical purity. Some might even be relatively unmoved by the idea of killing the odd animal. There may even be some who would like to eat tuna sandwiches every day of their lives if they hadn't made the kind and responsible decision not to.

So, if you wanted, you could even eat a whole plate of steak tartare one day (especially if it's made from responsibly sourced organic meat) and go back to being vegan the next day, and there's no *real* reason for anyone to give you a hard time about it. Black-and-white thinking about all this stuff can be very pernicious. Plus there's the fact that low-level meat farming, without the use of antibiotics, and organized around grazing and rotation, may be better in the long run than across-the-board veganism. The "Planetary Health Diet", recommended by Harvard professors in 2019, suggests that minimal meat and dairy eating may be healthier for humans and kinder to the planet than a more restrictive diet. Of course, this makes little difference to people who can't bear the idea of animals being killed for food. Still, even for these people, there surely has to be a huge difference between the cruelty and wastefulness of intensive beef farming, and the nature-respecting potential of humanely cared-for, grazing flocks of livestock.

So, if there are certain forms of food you adore, there can still be space in your life for you to enjoy them. The problem isn't you being a hypocrite; it's the possibility of other people taking their own confusion and bad feelings out on you. The occasional fuck-up is no biggie. And it may turn out that a little bit of what you fancy is the best environmental option anyhow.

Holidays

As with all the other stuff, judiciousness is everything. Also, to get things in perspective, flying – at least before COVID-19, which may reduce the numbers further – was only responsible for 2.5 percent of the world's carbon emissions. Still, we're talking about an actual emergency here, so every little helps. Plus, the flight industry has grown extremely fast and shows little sign of going out of business (global pandemics aside), so we certainly need to think about slowing it down a bit. If there are places you really want to go, however, you can still do it responsibly (assuming here that recreational flying will resume apace after the global lockdown is lifted). Carbon offsetting is a good solution and easily done – you can just log onto a website like Co2nsensus and they will organize it all for you. Of course, if everyone polluted like maniacs all the time, and then tried to type it away with a bit of offsetting, the system would soon collapse. Limiting your flights to treats and necessities (like visiting a dying friend or relative, perhaps – no one needs to begrudge you this) might be a better way to think of it.

Then there's the question of what you do on your holiday. Excessive tourism is destroying natural habitats, not to mention unsettling the human inhabitants of famous cities – who might be priced out of the rentals market by the explosion of Airbnbs. It's easy enough to research responsible tourism, and to plan your trip accordingly. And if all you've ever wanted is to fly a helicopter over the Grand Canyon, or to swim off the Maldives before they sink, perhaps you can just be aware that these things have effects and maybe you could counterbalance those in your own ways, such as by planting some trees, doing Veganuary, or giving up chemical hair dye (which can contain micropollutants that make their way into rivers).

Travel is amazing and it would seem unrealistic to tell people to stop doing it when the technology to get around is right there, and no longer very expensive. If you just do it less, do it responsibly and offset it, you can surely go wherever you like. Yay! The world is your oyster. (And, unlike almost all other seafood, farmed oysters are considered a sustainable food that's actually beneficial to the environment, not to mention rich in B vitamins, vitamin D and iron, so maybe you could even eat some for a treat sometime.)

Cars

Cars really are bad for the planet, but they are getting better. Still, if you *must* have one, there's the question of whether to get a new, greener one, or to keep going with the old one till it dies. The new car may burn less fuel, but it will be made from a load of plastic, rubber and metal (although, these days, some of this is likely to have been recycled).

Apparently, in most cases, the answer to this question is to go for the new car. Of course, constantly upgrading to the latest Tesla is hardly a responsible ecological choice. (Even though, as with fashion, this may mean that someone slightly less wealthy can replace their gas-guzzling Mercedes SUV with your old Model X, and so on, so that cars gradually become greener. Still, I believe it's been established that the so-called "trickle-down effect" isn't exactly reliable.)[34] But, unlike with so many other ecological decisions, this one is measurable numerically. According to a 20-year-old study titled "On the Road in 2020: A Life-Cycle Analysis of New Automobile Technologies", 75 percent of a car's carbon emissions are produced by the fuel it burns, with only 25 percent being down to production.[35] For a condensed account of this report, you can look up Green Car Report's excellent article online.[36] Here they explain how a newer car that does 40 miles per gallon will quickly make up for its own production against an older car that does 30 mpg. If you drive roughly 15,000 miles per year, the former will emit around 3.25 tons of carbon, whereas the latter will emit 4.35. The production of a new Mini produces around 6 tons of CO_2, meaning that if you look after your new car and keep it running for years, it will soon make up for itself in terms of emissions. Then there's the fact that more modern cars use less oil and require less maintenance and replacement of parts – not to mention the fact that an electric or hybrid car is likely to be even more fuel efficient than the hypothetical car in the example. So, if you can afford to replace your reliable noughties Volvo with a hybrid, definitely do it.

Of course, not everyone can simply splash out on a new eco-car, and this may begin to cause problems as new regulations aim to phase out the use of petrol and diesel. Perhaps in order to avoid the sorts of protests we have seen in France (by working drivers, or *gilets jaunes*, in response to increased fuel taxes), it will become necessary to offer grants to less wealthy drivers.

As with clothes, you can always hire a 1950s Maserati for the weekend if that's what turns you on.

Entertainment

Last but not least: entertainment, especially the forms of mass entertainment that involve screens. I don't simply mean actual phones, computers and TVs, which are all made from environmentally destructive materials – but which, unlike cars, shouldn't be replaced in the way that Apple wants you to each year: their production is far worse than their energy use. I also mean the film, TV and entertainment industries that make the stuff that goes *on* the screens. I mention it not to be a killjoy, much, but to point out that the solution to the problem of having a food-and-energy-consuming body isn't to stay at home and watch things. You might imagine that one way to avoid dressing up, being tempted to eat forbidden delicacies, and zip from one exotic location to another, might be to stay in, in your millions, and watch other people do it on your behalf. That way, Cate Blanchett gets to wear the dress, Daniel Craig gets to drive the car, and a load of C-listers get to go to the Australian Outback. Meanwhile you eat tempeh in your pyjamas and everyone's happy.

Sadly, it turns out that the entertainment industries are major polluters, in spite of the fact that so many of their major stars are great spokespeople for the environment. A 2007 study of air pollution in Hollywood found that, relative to its size, the film and TV industry was more polluting than aerospace manufacture, clothing and hotels. They were only really beaten by fuel refining. This is because most productions involve huge sets, generators for heat and light, semi-necessary minions (some of whom travel a lot), catering – which is rarely provided by local businesses. Because it's a hugely self-important industry, heavily populated by megalomaniacs, it's not unusual for the material demands of filmmakers to be uncompromisingly extravagant, resulting in overuse of energy, timber, clothes, human resources, plus the destruction of environments that need to be artificially transformed to support an illusion.

Some responsible filmmakers are well aware of this and try to do something about it. For example, Roland Emmerich insisted that *The Day After Tomorrow* should be a carbon-neutral film, and the Wachowskis not only took care of carbon emissions while making the *Matrix* sequels, they also used a non-profit organization to repurpose timber from their sets to build housing for low-income Mexican families, as well as recycling all of the steel. In total, 97.5 percent of the materials used in production were reused. So it isn't impossible to make films responsibly, you just have be inclined to do so.

The point in all this isn't to tell people to boycott the entertainment industry, as well as giving up all the normal stuff. It's more to say that opting out of real, embodied life isn't any kind of quick fix, as the Wachowskis have made a brilliant career out of explaining. Plugging ourselves into the Matrix is no solution to environmental issues, any more than the Voluntary Human Extinction Movement is. You can see the insane practicality of it, but it's also cynical and foul. Far better to keep finding interesting forms of enjoyment, probably involving other people, even if they sometimes produce a little waste. As my friend reminded me when I was told her I'd renounced dating in order to get more involved in activism: "Love is carbon neutral." Enjoying real life is not a crime. Which is why the next chapter will focus on environmental actions you can take that aim to enrich your life, rather than deplete it.

PLEASURE: CHAPTER OVERVIEW

1. Don't aspire to give up all the things you like, even if some of them are a bit naughty.
2. Never feel guilty about your own existence. While it may be true that human beings have done a great deal of damage, we're also a truly incredible phenomenon. We deserve to be saved from extinction every bit as much as snow leopards.
3. Learn to sew: that way you will always do well in charity shops.
4. If you like your parents' clothes, "borrow" them.
5. Opting out of enjoyment can make people bitter. Don't let that be you.
6. Please don't bid against me on eBay.
7. Eat meat, go on holiday, get a new car. Sort of.
8. Don't let environmentalism frighten you into a lacklustre existence. A responsible life can still be utterly brilliant.

8
DO THE RIGHT THING (WHATEVER THAT IS)

"We don't need a handful of people doing zero waste perfectly; we need millions of people doing it imperfectly."
Anne-Marie Bonneau, @zerowastechef

It can be horribly confusing trying to work out the best way to make a difference – should you give up your car or go vegan? Fly less or buy fewer clothes? Does domestic recycling help at all? Or are small personal consumer choices like these beside the point: would you be better off focusing on lobbying governments and businesses, or even standing for parliament? Some sources will offer very definite opinions, such as: have one less child (save 58.6 tonnes CO_2), live car-free (save 2.4 tonnes CO_2), avoid aeroplanes (save 1.6 tonnes CO_2), and eat a plant-based diet (save 0.8 tonnes CO_2).[37]

However, while lists like this may be appealing in their simplicity, they ignore the complexities of human life and social systems (as we saw in Chapter 6). Even if you cross off the first suggestion for being somewhat facile, the order of the next three also turns out to be up for question. Some people argue that cars are worse for the planet than meat-eating, while others say the opposite. Also, what does "plant-based" actually mean? Going vegan, or only eating meat sometimes? How much would you have to fly in the first place in order to save 1.6 tonnes of CO_2? These calculations will have been reached by pooling a huge amount of data and coming up with an overall figure, but the pool itself will vary from study to study; for example, some will only take into account carbon emissions, while others also focus on the use of plastics, pesticides and other greenhouse gases. Still, you might think that if you give up your car, stop flying *and* go vegan you can't fail – but then someone will come along and tell you that responsible, organic flexitarianism is better for the planet as it's more nutritionally viable, and that grazing cows help with biodiversity and the prevention of soil erosion. Or that factory workers will suffer, causing social unrest, if we all stop buying cars. Or perhaps that temperatures actually *rose* after the Icelandic volcano eruption at Eyjafjallajökull in 2010 – and its subsequent ten-day flight ban – due to the lack of aeroplane contrails shielding the earth from sunlight.

An article published in *Nature* magazine argues for "a synergistic combination of measures" – i.e. making a variety of changes and keeping an eye on things as the consequences unfold.[38] If everyone went vegan at the same time, it's true that the adjustment period would be a mess. By being hesitant, open-minded and unsure, you may even be more helpful, or at least *as* helpful, as someone who is certain and dogmatic. Big, monolithic solutions are unlikely to be the answer – although massively reducing the use of fossil fuels is the one thing all climate scientists agree on. Still, how you go about it is up to you. If you have a close relative who lives abroad, perhaps you won't want to give up flying altogether, in which case you can cycle everywhere, change your diet, or alter the way you heat your house.

The purpose of this book isn't to tell everyone to give up everything and become a full-time eco-warrior (although I fantasize about doing this all the time). But it can obviously be extremely helpful to have *some* idea of where to focus your energy. The very blurriness around the choices you're being asked to make can be anxiety-provoking. This chapter will make suggestions about practical changes you can make in all areas of life, alongside more creative, poetic actions that might also have far-reaching consequences. It's not about obsessing over the details, counting each milligram of carbon you do or don't emit – please keep breathing! – but adopting an open-minded, sensitive and responsible approach to the world and everything in it.

If you're reading this, it's presumably because you're already hyper-aware of the scale of the problem and hardly need to be bossed into action. So please don't take the following list as a set of diktats designed to make you feel even worse about yourself. It's more of a data pool for you, and me, to check in with. I'm sure there are plenty of things I'll forget, or omit out of ignorance, not to mention the things that will inevitably change. As we've seen, even in a short space of time, almonds can go from good to evil, and new food technologies might make the question of veganism versus meat-eating less pertinent (although synthetic meat isn't entirely cruelty-free as it requires donor animals, so won't be acceptable to all animal rights supporters). The overarching aim of the list is to make the planet a priority in as many viable, enjoyable and effective ways as we can. Planet-friendliness is a mindset rather than a persecutory list of commands. Once you start to see things a certain way, it becomes easier and easier, to the point where it doesn't feel like "making sacrifices" or "giving things up'" – more "Why would you even want to go for a helicopter ride when you could do something non-toxic, and frankly much cooler, like admiring a tree instead?" It's a case of enjoying the love you feel for our world.

Home

In the UK, the heating of buildings accounts for more than a third of our greenhouse gas emissions. More forward-looking countries, like the Netherlands, have effective schemes to reduce emissions, such as ensuring all new buildings will be near enough to carbon neutral by 2021. Whether or not you live in a "green" country, switching from gas to sustainable electricity at home – or in your workplace if you can persuade your employers – makes a big difference. This might involve serious outlay if you install your own solar panels. Over time they may very well pay for themselves, though, especially if you sell your surplus energy back to the national grid. Some governments and councils have financial schemes to encourage people to do this – although there are other twits, like Donald Trump, who are trying to get people back into burning fuel "for the sake of the economy". Luckily, this is what's known as a "zombie idea" – i.e. it ought to be dead but some malign force is keeping it going, and happily Americans generally don't seem to be listening.

- If, like me, you can't currently afford solar panels, you can at least switch to a clean energy provider and try to choose electricity over gas as much as possible. For instance, you could use an electric blanket before bedtime in winter, rather than putting the heating on. Also, don't be embarrassed about being stingy – wear bedsocks at night and big jumpers in the daytime. Or get heavy, draft-excluding curtains. You can cordon off areas of your house with curtains (bought on eBay if you're so inclined) so you can heat the smallest areas necessary. I've just installed a really cheap curtain along my bannister and it works brilliantly.
- Double-glazing can help a lot but is another big outlay, and not much help to renters who may have very little say in this sort of thing. However, building regulations in many countries are now pushing landlords toward more responsible approaches to energy conservation, so you might get them without even having to ask. If there's no way for you to acquire more-insulating doors and windows, you can seal off drafts with draft-excluding tape. Draft excluders are great, if kitsch.
- Sometimes simpler solutions may actually have advantages over the big, expensive ones. For instance, wearing extra layers of clothing can immediately reduce your gas emissions without all the wood, metal, glass and plastic

involved in installing panels or new windows. If, at the worst possible estimate, we may have only a couple of years to go before cataclysmic weather events start to make life on this planet way more difficult, it might make more sense to go for these immediate, no-outlay solutions, rather than changes that use more resources, so take longer to offset. As ever, perhaps the best thing is for different people to do different things, according to their means, and for everyone to remember that poorer people tend to have smaller carbon footprints anyway, so should *never* be made to feel bad for not making big, visible gestures.

- Cleaning products are another area where it's possible to make more environmentally conscious choices, although of course the eco brands tend to be more expensive. If you're on the broke side, you can get a very long way by foregoing chemical-packed supermarket brands in favour of vinegar, soap, salt and baking soda. Even water alone can get things clean, and pre-soaking stove tops and gungy saucepans can make abrasive powders completely unnecessary. Books like Karen Logan's *Clean House, Clean Planet* are full of ideas about how to rid yourself of toxic chemicals without resigning yourself to a life of grime.
- Keep your appliances for as long as you can, and dispose of them responsibly when they die. Replacing a fridge or washing machine just because you feel like a nicer one is scoundrelly. The earth is not built to accommodate this sort of mischief.
- Switch off the tap while you brush your teeth and take showers instead of baths as they use less water – especially if you're quick. People often forget that domestic water is linked to CO_2 emissions, and not just the hot stuff. In California, 20 percent of the state's electricity is used to treat and pump water, not to mention 30 percent of their natural gas, after they've deducted the amount used by power plants. Water usage often tends to get forgotten in discussions of the wasting of resources. Less energy used for

pumping, and fewer chemicals used in treatment, means a lower carbon footprint; chucking unused water down the drain is an area you can mark as "needs improvement".

- If you have pets, aim to do so responsibly. Don't buy them from profit-driven, unethical breeders, but try, if you can, to get a rescue pet. For reasons too complicated to go into here, I have a vegetarian but massively unethically bought dog (don't ask!) and a meat-eating rescue cat. I totally see people's point when they say that keeping domestic animals, especially in cities, is fundamentally wrong. Still, what are you going to do if you already have some creatures, and love them? It's not like you can release them into the wild. Dogs seem to do really well on a vegetarian diet, though. Cats not so much – although that doesn't seem to stop people manufacturing vegetarian cat food.

In the kitchen

As we've possibly agreed by now (I hope), eating less, or no, fish, meat and dairy is a given. If you decide that flexitarianism is the one for you, aim for ethically produced organic meat and sustainably caught fish, and don't eat very much of it. Of course organic stuff is more expensive, but that will easily be balanced out by the money saved on eating more plant-based meals in between. In July 2019, a group of researchers at Oxford University calculated that going vegan could reduce your carbon footprint by 73 percent.[39] While numbers like this are unreliable – what if you kept driving an SUV and lived on vegan takeaways in polystyrene boxes? – it still gets across the idea that changing your diet can make a substantial difference. There's also the fact that these things have a cumulative effect: the more people cut meat, fish and dairy from their diets, the more shops and supermarkets have to cater for them, which makes a plant-based diet seem easier and more acceptable to people who hadn't considered it before. On top of this, there's the power of suggestion – if everyone has the sense that eating less meat is becoming the norm, then we are more likely to join in with it. (I say "we", but I know that some people have been doing it consistently for decades and might even feel a bit miffed when their local shop is suddenly sold out of oat milk and tempeh throughout the whole of pesky Veganuary.) So, when you go vegan, vegetarian or flexitarian, it's not just about your own carbon footprint, but also the footprints of anyone you manage to "infect" (if that's not too tasteless a metaphor these days).

While it's great that supermarkets have jumped on the vegan bandwagon, don't give up on smaller health food shops who've been putting in the groundwork, making veganism and vegetarianism possible for decades. Apart from being heroic, they also tend to sell more hard-core brands that package their products in glass, card and waxed paper, rather than plastic.

Whatever kind of diet you eat, try not to waste food. In the UK, between one-fifth and a quarter of food produced is wasted. Seventy percent of UK food waste is produced by households, while the rest is divided between retail, manufacture and hospitality (with retail being by far the least wasteful, at 3 percent). This means that in the UK we produce 6.6 million tonnes of domestic food waste – and this figure excludes the "inedible parts" like bones and stones; all of it is food that could otherwise have been eaten. (America is even worse, wasting up to 50 percent of the food it produces.) The food wasted in the UK in 2018 had a value of £19 billion and is calculated to have produced 25 million tonnes of greenhouse gas.[40] Council food waste recycling schemes are helping to slow things down a bit, but not nearly enough. It would be better if people just bought less food and made sure they ate it before getting new stuff. Soups and stews are brilliant for using up odds and ends, and everybody knows that the best meals happen by surprise when you're using up your leftovers.

- Don't overfill your kettle as this wastes electricity.
- Buy local as much as you can. It makes no sense to go vegan and then live on exotic fruits from the other side of

the world. Veg box deliveries can be great for sorting this side of things out for you – companies work with farmers to provide an interesting variety of seasonal fruit and veg, and they also tend to recycle, employ eco-friendly delivery methods and avoid plastic. It's also really fun when the box turns up and you get to meet your vegetables.

- Go plastic-free. This will may very well involve changing the way you shop – supermarkets are terrible. Also, bear in mind that "plastic-free" needn't mean throwing out really useful things you already own, like Tupperware boxes or KeepCups. This won't save the planet, it will just add to landfill. It may also mean you have fewer places to keep your leftovers fresh. You only need to avoid buying new things that come covered in unnecessary plastic, like cucumbers or bagged lettuce. If you have an old-fashioned greengrocer nearby, they often turn out to be cheaper than supermarkets.
- While we're on the subject of bagged lettuce – mass produced, plastic-wrapped sandwiches are also bad news. Some of us still remember the days before these ubiquitous menaces were considered normal, so it can't be the case that life can't go on without them.

Garden

Do whatever you can for bugs and bees – make your garden as biodiverse as possible. Even window boxes full of lavender or bee-friendly flower mixes are great; and leave the dandelions alone as these are one of the most important early flowering plants to grow in countries like the UK, providing pollinating insects such as bees as well as beetles and hoverflies with vital resources in the spring. If you are a fastidious gardener, try to keep areas of your garden undisturbed so creatures can build stable habitats. Woodpiles are lovely habitats for fungi and beetles, and indigenous weeds are better for wildlife than fussy, exotic flowers –

although of course you can always include both. Some "weeds" such as comfrey, cow parsley and buttercups are as beautiful as any "on purpose" plant from a nursery.

- If you have space, plant an air-cleansing tree. Some species are particularly good at absorbing pollution and generating oxygen: Mediterranean hackberry, field elm, common ash, littleleaf and largeleaf linden, Norway maple, turkey oak, and gingko are the best. See the LifeGate website for more information about how each of these trees can not only convert CO_2 into biomass, but also trap benzene, nitrogen oxide and dioxins in their cuticles and plant hairs, thereby cleansing the air of pollutants.[41]
- Plant fast-growing fruit and nut trees in case of the apocalypse. (Don't mean to stress you.)
- Lawns are nice for sitting on, but not much else. I have one and am ashamed of it, and plan to phase it out. They give nothing much either to the atmosphere or to wildlife. Constant mowing means that insects can't make stable homes there. There's also the fact that lawn maintenance encourages a kind of horti-fascism: if you look at standard "lawn-care" products, they're all about killing off anything that doesn't have a place on a professional golf course. Oh my God, what if a bit of moss caused your ball to bounce differently – and as for dandelions and daisies, ugh! Lawns are all about domination over nature – Las Vegas is full of them. Lawn space could so easily be given over to wildflower meadows, vegetable patches or trees. Down with lawns!
- Grow your own herbs and veg if you have the space, obvs. And if you must have cut flowers, maybe grow them yourself rather than buying the toxic blooms you see in supermarkets.
- If you don't have a garden but your mum does, tell her you won't even consider making her a grandmother until she gives up her lawn and lets some proper plants grow.

Clothes

We talked about clothes in the last chapter. To summarize: no more fast-fashion and rewear the fuck out of everything you own. If you think second-hand clothes are no fun, watch the web TV series *Sex Education*: the costumes are also a brilliant example of non-wasteful TV production.

Cosmetics

There are plenty of great vegan, cruelty-free brands these days, but they still tend to come in a load of difficult-to-recycle plastic. As with so many other things, less is more. If you love make-up and skincare stuff, go for the most ethically produced stuff you can afford and try not to buy a ton of it.

- If you live with other make-up wearers, perhaps you can do swaps?
- Apparently foundations are one of the most wasted cosmetics because, in search of the perfect product, we so often buy shades and textures that don't suit us. Three ways around this are (i) do a ton of research to find the most recommended, easiest-to-wear brands, (ii) try a sample and look at it in daylight before you buy it, and (iii) if you find one you like, stick with it. Don't end up with eight different foundations going stale in your make-up box just because you fell for advertising/felt depressed in a department store/had some nerve-wracking events to go to. (Note to self, this means you.)
- Don't follow Instagrammers who make you want to buy unnecessary crap. Being creative within limited means is obviously infinitely cooler.
- Plastic glitter is terrible: it creates microplastics, which are bad for plankton, fish, sea mammals and birds. Also watch out for fake biodegradable glitter. Since people have become far more aware of the problems caused by microplastics, glitter manufacturers have regrouped and now offer

things like "compostable", "PLA" and "cellulose acetate" glitter. None of these dissolve or biodegrade under natural conditions – they require special chemical treatments which they simply won't receive if they get flushed down the sink. Only products made from "Bioglitter" (a trademarked substance) are made from plant cellulose and can be metabolized by microorganisms. If you can't find out from the packaging what kind of glitter it is, don't buy it.

- Avoid wasteful make-up removal wipes and never, ever flush them down the toilet.
- Learn how to make your own face, hand and body lotions. There are loads of online tutorials and recipes. Lotions are just emulsified oils, so can easily be put together at home with no animal testing, dodgy packaging or unnecessary chemicals. You can add your own scented oils to make them smell nice, which also have antiseptic, relaxing or enlivening qualities (tea tree, lavender and peppermint respectively).

Transport

By giving up your car, you will reduce your carbon footprint by an average of 4.6 tonnes of CO_2 per year. This calculation is based on a gas vehicle that does 22mpg and travels 11,500 miles per year, so you can increase or decrease that figure according to what kind of car you'd be giving up.[42]

- Ride a bicycle, it's lovely. I've been riding bikes around London for thirty years and haven't been killed yet. As more cities get proper cycle networks, horrible accidents become even less likely. Also, cycling keeps you fit without having to do boring, artificial, non-functional, cosmetically driven gym stuff.

9. Public transport is completely brilliant, as is walking. Let's hope pandemics don't keep us off tubes and buses in the future. Presumably gloves and masks are about to become a good deal more popular…

- If you live in a town or city, schemes like Zipcar are great. They tend to have loads of low-emission vehicles, and sharing cars means you don't have to manufacture so many in the first place.

At Work

If you come up with any brilliant environmentally friendly schemes for your workplace, try to introduce them in the nicest way you can. Some ideas, like lowering the thermostat a degree or two, will save your company money so should be an easy sell. If you want to do something bigger, like persuade them to install solar panels, perhaps you could offer to help fundraise by organizing a sponsored walk or something. Haranguing people and making them feel guilty almost never goes well. Part of being an effective environmentalist involves making it appear as easy and enjoyable as possible. Of course, there may be times when it's appropriate to glue yourself to a building or otherwise cause a nuisance, but cajoling colleagues and bosses into better environmental habits might be better off handled sweetly and tactfully. Try to take as much responsibility for the changes as you can – like offering to buy and make labels for bins to separate rubbish – and avoid bad-vibe, snippy notes.

Treat workplace environmentalism as a stealth takeover mission. Even though people might say that things like recycling have minimal effects, perhaps you can see it as more of a "hearts and minds" matter. For instance, if your boss notices that the staff all seem to care for the planet, perhaps it will embarrass him or her into switching to a more eco-friendly car.

Investments

If you have savings or buy shares, be very careful about whom you give your money to. BP and Exxon do not deserve to borrow your cash, nor do any of the other top 20 global polluters. And of course you wouldn't be fooled by their nascent sustainability efforts. They don't need your money to fund these – they can stop giving their CEOs bonuses until they've come up with *much* better ways of conducting business.

Look for ethical investment funds, rather than just accepting whatever ISA or savings option your bank offers. The Charity Commission in the UK discovered that even charities have tended to opt for non-ethical forms of investment, with the idea that their prime responsibility was to raise as much money as possible for their cause, even if this meant investing in businesses that ultimately went against the aims of the charity. This obviously makes little sense – especially not these days when ethical investments may even turn out to bring better dividends. In the UK, the publicly owned pensions provider, NEST (National Employment Savings Trust), found that their "climate-aware" funds did particularly well, delivering an unusually brilliant return of 116 percent over seven years, beating less-saintly investments.

Activism

This can mean any number of different activities, from large-scale disruptive actions to tiny conversations. Activism can be as literal or artistic as you'd like it to be. You can join organizations that are already up and running, or you can do things your own way. If you think XR are tin-eared, clueless white people, you could join Wretched of the Earth, a climate justice collective of indigenous and BAME organizations. Also, if you're worried about class issues around climate activism – i.e. middle-class twits telling blue-collar workers to give up all their hard-earned pleasures, while trying to close down the "polluting" industries these same workers work for – there's a brilliant article by Tristan Cross on the Dazed website: "Don't be Fooled into Thinking That Climate Activism is Just for Poshos" (17 October 2019).[43] In it, he explains how the wealthy right exploit this line of thinking in order to divide and conquer, as they're actually the ones who stand to benefit from continued business as usual. They definitely don't give a toss about low-paid workers beyond their lovely exploitability. Calling XR a middle-class Doomsday cult does little to refute climate science. The problem is that this strategy works really well – everyone loves to hate privileged hand-wringers. We even hate ourselves.

You can decide what you'd like the purpose of your activism to be – say, to put pressure on politicians, or to help ordinary people to feel better. If it's the former, you might write letters (especially open, public, multi-signature ones), organize petitions (while being aware that "clicktivism" is only really useful for collecting lists of sympathizers in order to motivate them to do something actually useful, even if that's just to donate money), or get involved in local or national politics. If it's the latter – you want to help others to feel better – perhaps you could set up a local group to work on environmental ideas for your area, like guerrilla gardening (rewilding disused spaces), or persuading local food shops to donate excess stock to homelessness charities.

Or start a "Climate Café" for fellow sufferers of eco-anxiety. The website www.climateandmind.org outlines the principles of Climate Cafés: namely to give people spaces in which to congregate and discuss their ideas and fears, with the ultimate aim of encouraging them to feel empowered to speak out and make changes. The website also gives advice on identifying goals, structuring meetings and offering psychological support to people who are really suffering.

Other ideas for activism might be to have "consciousness raising" market stalls, paint murals (like the big one of Greta in San Francisco), campaign for a beef-free canteen, write brilliant poetry, make excellent films (even on your phone, for YouTube), organize a catwalk show with your local charity shop, or absolutely anything you can think of that lets other people know about climate change in a friendly, engaging way. Maybe pick one point on this list and see what you can do with it. If everyone in Norwich, say, refrained from overfilling their kettles, our air would be that little bit cleaner. The more people give a damn, the better. Even governments and huge, powerful organizations can be swayed if they feel (or fear) that there are enough of us out there to dislodge them if they don't shape up. We have to make them see that they can't sit back and watch the planet suffer for the sake of their own comfort. The more we show we care, the more difficult it becomes to ignore us – especially now that the scientific community is openly and outspokenly on our side.

ACTION: CHAPTER OVERVIEW

1. Drive fewer cars, eat more plants. However, recognize that things are complicated and that nothing is purely black and white.
2. On the subject of black and white, try to be racially sensitive about your activism. Also, if you're white, be aware that arguments about giving up materialism might mean something different to groups of people who have historically been excluded from owning property. One rule for Cardi B, another for me, and that's fine.
3. The world is exciting enough – it doesn't need glitter-coating.
4. Replace lawns with almost any other kind of plant, especially wildflowers and trees.
5. Try not to lend money to polluters.
6. Be really charming when trying to persuade colleagues and/or employers to make your workplace greener. No one wants to be the subject of behind-the-back eye rolling.
7. Don't be tricked out of activism by people who don't have the planet's best interests at heart.
8. Please don't let lists like this one add to your worries. I'm sure you're already doing most of it anyway.

9
HOPE IS ALLOWED: DON'T STOP BELIEVING

> *"Hope is the thing with feathers*
> *That perches in the soul –*
> *And sings the tunes without the words –*
> *And never stops – at all ..."*
>
> Emily Dickinson, poet

Within the world of climate activism, hope is an understandably suspect sentiment. Greta Thunberg famously doesn't want your hope and XR spokespeople seem to have rejected the word in favour of "courage". The risk with "hope" is that it might sound somewhat passive. But can there be active forms of hope? And how might hope and courage be connected?

In this chapter we'll look at the story of Plenty Coups, a long-dead American Indian chief whose thinking has been very important to the climate psychology movement. We'll also consider the difficulty, and perhaps the importance, of bearing uncertainty – which takes us right back to where we started, with anxiety. By bringing together some of the ideas in this book, how might it be possible to face our uncertain futures with courage and hope, and not get lost in fear of things that might or might not happen?

Radical Hope

You may have clocked that in people's thinking around the climate emergency, the word "radical" comes up a lot. There's radical ethics, radical hope and sometimes radical politics.

What does it all mean? The word suggests a huge change or overhaul, but can also be used simply to mean "extreme". While you might not think it was possible to save the world by being extremely hopeful, not everyone has given up on the idea of hope as an agent for change.

In 2006, the American professor of philosophy Jonathan Lear published his book *Radical Hope: Ethics in the face of cultural devastation*. In it, he describes how, toward the end of his life, the Crow chief Plenty Coups told his life story to a white man called Frank B. Linderman. Plenty Coups was around eighty years old at the time, and had lived through the war for the West, fighting alongside General Custer – against the Cheyenne and Arapaho tribes – in the Battle of the Little Bighorn. One of the things that struck Jonathan Lear in Linderman's written account was Plenty Coups's idea (only mentioned in an author's note near the end of the book) that, after white people took over, nothing more of interest happened throughout the rest of his life. He could hardly get across all the amazing stuff that had gone on previously – raids (or coups), losses, celebrations – but then it all went dead. By white American standards, quite a lot happened: he won prizes for agriculture, took on the Senate over land rights (and won), went to visit George Washington's house, and donated his home to the nation. But by the standards of the culture he grew up in, these were not really *things*: he was just trying to make the best of an unprecedentedly abominable situation. In his previous life, and the life of his tribe, existence was organized around catching wild buffalo and planning raids on the neighbours, collecting as many scalps as possible. Ethical life centred on the idea of honour and never backing down; once you'd stated your intention to fight, you either had to win or die. Anything else would be inconceivably shameful. This sort of thinking, not to mention your claim to the land you lived in, made you totally unacceptable to Christian colonizers, who would criticize your savage barbarousness while also planning to kill you. If you were the type to stick to your guns (or less efficient weapons) and fight to the death, this new type of enemy would clearly wipe you and your people off the face of the earth. So how might it be possible to survive without being a backtracking coward? The only way to do it would be to forget pretty much every principle you had ever held dear, and to hope that a new set of principles would be able to fill that void in the future. If you could just hold on to life, somehow you'd be creating space for a new set of ethical possibilities.

Lear acknowledges the possibility of being "blinded" by hope. He tells us: "We use the term 'Pollyanna' pejoratively to designate someone whose hopefulness depends on averting her gaze from devastating reality. Indeed, we sometimes suspect that a person's hopefulness is a strategy for averting her gaze."[44] But this clearly wasn't Plenty Coups's case: the choices he made in the face of attack were nothing if not purposeful, even if his purpose might only have been able to unfold or reveal itself later. He was simply betting on the idea of existing, even if that new existence contained absolutely none of the things that had given his old existence meaning.

So why do a bunch of peace-loving climate psychologists have such a thing for an American Indian warlord who sold out his neighbours to fight alongside white people in the Wild West, without even being altogether sure why he was doing it? To borrow words from the "Climate Change and Radical Hope" issue of *The Psychotherapist*: "The hope was 'radical' because the Crow needed to be true to themselves, and yet a transformation of their culture was required beyond their imaginings."[45] They couldn't possibly know in advance where this transformation might take them, nor what kind of people it would turn them into, but they chose to follow Plenty Coups, and to take the leap. The things that had given their lives meaning, and around which they had built their identities, were snatched away from them; their way of life was made illegal. But instead of becoming hopeless and depressed, they found ways to carry on and, eventually, to produce new forms of meaningfulness. In the place of the old warrior–hunter ideology, Plenty Coups began to develop a philosophy of education and cooperation – both learning from white people, and teaching them the very particular knowledge of the Crow (hence winning prizes at agricultural shows). The fact that his house is now a museum, situated within the Plenty Coups State Park in Montana, is perhaps proof of how well this went.

The reason this matters to climate psychology people is that it shows there are ways to bear seemingly unthinkable losses. Although we can't know how the climate crisis will pan out, we have to contend with the possibility that it will involve the destruction of cities and landmasses, the permanent extinction of thousands of species, and large-scale societal breakdown. (Remember Jem Bendell's apocalyptic predictions in Chapter 3?) Still, there's also the chance that science will save us. Or that climate change activists are just being alarmist. Or that we will all pull together to avert catastrophe. Or even something in between, perhaps only *really* affecting parts of the world with less temperate climates, meaning that the rest of us can just keep going so long as we don't think about it too much. (Many would agree that we're at this point right now.) The trouble is that we can't possibly know. Either we're going to hell in a handcart or we aren't, and different people will tell you different things about which, why and when. Climate change science is utterly convincing, but the fact that people understand it much better also *sometimes* means they are doing more to counteract the alarming effects that are already well underway. (See Chapter 10 for some encouraging examples.)

As we said right at the beginning, anxiety is different from fear in that it suggests an uncertain outcome. The current state of the world is nothing if not uncertain, and anyone who tells you otherwise is deluded. Therefore, perhaps, anxiety is the psychological state *du jour*, with statistics suggesting that around 3 million people in the UK are suffering from a diagnosed anxiety disorder, which leaves anybody to guess how many of us might be suffering from an undiagnosed one. We have every reason to feel anxious. But this also means we have the option to face our uncertain future with courage and with hope.

In therapy, people sometimes say they don't like to hope because it only leads to disappointment. But the killing off of hope in oneself can be quite a cowardly and controlling act. It may keep you in a state of stasis, meaning you don't need to try. It can also contribute to bringing about the very situation you were supposedly trying to avoid. If you've given up hope of ever falling in love again, for example, why would you bother to take steps that would make it more likely to happen?

So, while sitting back and hoping it'll all be alright might be naive and complacent – and if everyone did it, we'd be screwed – if we *stop* hoping, we might also be covertly inviting catastrophe. The trick is somehow to make your hope *active*: to use your hope as an engine for change.

Active hope offers a way through our difficulties, however unmapped and unpredictable, and this kind of hope requires courage. With active hope, you may see the point in protesting, in studying for your science exams, in producing an ethical clothing line, in standing up for your opinions, or in having a baby and bringing it up in the best way you can. All of these things are brave acts. You can't say for sure how they'll pan out, whether they'll go well or badly, but trying has got to be better for your sanity than not trying. At the very least it gives you something to do. At most, you may actually make a huge difference.

What Kills Hope?

Here are a couple of examples of the sorts of habits and emotions that can dash hope, followed by a more optimistic observation – about the constructive ways in which we can harness these …

Negative ruminations

Perhaps one of the biggest dashers of hope – and contributors to anxiety and depression – are negative ruminations. This is where your mind takes hold of something, turning it over and over, and refuses to let it go. If you are amongst those of us who are kept awake fretting about the damage being done to the world, it can feel impossible to separate yourself out from your negative thoughts. Why should you? These bad things are really happening. How can anyone just doze through them? It can feel as though every poisoned bug, every bleached coral, every krill-starved whale is all at once there in bed with you, at the same time as being a million miles away, beyond your capacity to connect. And it can feel like a form of cowardice to look away, to allow for distractions – including essential ones, like sleep. But if these thoughts are getting in the way of your capacity for hope then they definitely aren't helping you or the planet. Try to remember that hope is allowed, and that getting lost in negative feedback loops only serves those people who don't want change.

Anger

Anger, too, can be a hope-killer. It's possible to become so fixated on the self-serving blindness of people in power that you end up feeling bitter and paralysed: your own weakness in the face of their strength makes you dislike or discount yourself. Or give up on humanity as a lost cause. But perhaps it's possible to remind ourselves that, in the same way as we are all connected to the planet – through the food we eat, the air we breathe, the water we drink – we are also connected to each other. If enough of us let it be known that we don't accept dishonest economic arguments around fossil fuel subsidies or fracking, then the untouchable people become touchable – we can affect them. We can withdraw our votes from politicians who support anti-environmental practices; we can withdraw our money from banks and insurance companies that support dodgy businesses; we can withdraw our custom from retailers who sell irresponsibly sourced products, and we can refuse to engage with institutions, such as galleries, museums and theatres, whose funding depends on oil money. If we don't appear to have much power as individuals, we certainly have power as a mass. By acting in the ways we think are right, and by being open to communicating our ideas about why it matters – through speech, writing and/or protest – we can apply an enormous amount of pressure upwards. (Or downwards – why should *we* look up to *them*?!) We can make governments and businesses seem small by comparison.

Due to the speed of communication, good ideas can spread quickly, mobilizing people, inspiring them. And ideas are free. Your good idea might even take off without you particularly having to do anything to promote it, like Greta Thunberg's school strikes. It might spread in the face of huge resistance, like Jem Bendell's "Deep Adaptation" paper. Or you might successfully employ the most carefully researched communication tactics and strategic protest styles, like XR, to create a worldwide movement. Alternatively, you might be

"Over the last year, I have become increasingly aware of how rapidly climate and global warming are happening – and the knock-on effects on animal species, habitats, weather patterns, mass migration from drought areas, etc. I have also noticed that for the first time in my life, I am waking up several times a night, worrying about melting glaciers, burning Arctic forests, methane and carbon emissions, plastic pollution of the oceans – and an overwhelming feeling that capitalism, corporate greed for resources and an utter disregard for nature are just ruining the planet which sustains all our lives. It makes me feel deeply sad, and alarmed and frightened about what the future holds for our planet."

Alison Ryder-Cook, university graduate

someone who can see the excellence of other people's ideas, and take them up on it, adding your body, your time or your thinking, helping to turn their grain of sand into a pearl. But whatever you do, it's bound to involve a certain amount of active hope and courage.

Saying the Unsayable

Perhaps one of the most surprising, and hope-giving, phenomena of the last few years is the capacity of people to hear, and to bear, the truth. After decades of nervousness/doubt/dishonesty hampering the discussion around the realities of climate change, the truth – in so far as we are able to know it – has finally been allowed to emerge. Unfortunately, this is perhaps because it has become impossible to deny: changes that had been predicted for decades are visibly taking place. Still, it wasn't, and isn't, easy for people to say it. Thunberg, Bendell and XR have all been subject to attacks on their credibility, being called anything from mentally ill to unprofessional, to batty or counterproductive. The reasons put forward for staying silent seem to cover all bases: what if you're wrong? What if you're right and you scare people? What if no one believes you? The risks of speaking out might seem too big. In my own profession, nervousness about expressing *any* opinion, or of losing one's position of neutrality, has sometimes resulted in a kind of frozen apathy around political matters. Therapists don't want to admit that they're scared, or that they might have fallen for hype, or they fear that their clients will disagree with them and leave. (Many of the psychotherapeutic community also, in my opinion, have an abominable attitude toward vegetarianism, often considering it "immature" or merely "a protest against something else". Can we cut that out now, please?) Nonetheless, the truth has somehow filtered through in all areas of society and people are showing every sign of preferring it that way. At least from here we can make informed choices.

For children and teenagers in particular, being shielded from the truth has evidently been infuriating. Patronizing young people by claiming that lies and obfuscations are somehow "for their own good" has been blown apart as a strategy. They have excellent ways of spreading information among themselves, and if the adults in their environment won't be straight with them, then they'll look elsewhere for better people to exchange ideas with. Schoolkids have shown incredible bravery in facing down condescending politicians, police, teachers and angry counter-protesters in order to get their message across. It may be the case that their voices have had the most power of all: what kind of person can blot out the hopes and fears of children who are faced with solving a gargantuan problem they had no hand in creating? While Greta Thunberg might say, "I don't want your hope", she demonstrates her own radical, active hope in everything she does. And in so doing, she makes way for other people to follow. Now that a whole generation of children have found their voices regarding climate change, it will surely be impossible to silence them.

Active Hope

Anxiety, uncertainty, hope and courage might initially sound like an odd collection of feeling states or attitudes. Do they contradict, cancel and undermine each other? Or might they actually fit together very well?

As we saw in the opening pages of this book, anxiety – according to its second definition – is, in itself, a form of hope: "I'm anxious to do the right thing by the planet." But this type of hope is linked to worry, and worry is fed by situations with uncertain outcomes, which may require courage to face. There's also the fact that allowing for uncertainty *at all* requires a degree of audacity. It takes strength to admit that there are things you don't or can't know. It's easy to get stuck behind a fixed idea, like a curtain that shields you from the world. So if your worries make you uncertain but also curious, and lead you to ask serious questions about what you might be able to do, then you are already being brave. You are opening yourself up to the possibility that your actions might fall short, or might turn out to be the wrong ones, but the fact that you are prepared to think and to try puts you way ahead of people who are too scared to look beyond the familiar. If you stay open to the world, and to other voices in it, you can keep listening and responding to different ideas, fine-tuning your thoughts and actions as the future rolls out before you.

The reason that Plenty Coups has become a legendary, inspirational figure of hope isn't because he knew what he was doing and steamrollered ahead. It's precisely because he had *no idea* where his actions might lead. As Jonathan Lear explains: "[A]t a time of cultural devastation such as Plenty Coups faced, the risks include not only malnutrition, starvation, disease, defeat, and confinement; they include *loss of concepts*."[46] This is where we find a direct analogy with many of the fears we face today. We are told the planet cannot provide for us at our current rate of consumption. If things don't change drastically, we will be faced with the problems listed above – or, equally disturbingly, we will carry on consuming as we are, leaving other people to face these problems. (And if you need to be reminded why this is a problem for *everyone*, watch Bong Joon-ho's Oscar-winning movie, *Parasite*.) But in either case we may suddenly find ourselves in a world we barely recognize, either due to our own lack of necessities, or to major social upheavals triggered by other people's. Societies may alter beyond recognition. Our current ideals and aspirations, or those we have for our children, may have no place in this yet-to-exist world. And yet, like Plenty Coups, we might choose to see what new meanings we could make there. Rather than being so frightened of the future that we cease to invest in it, we can choose to have a hand in shaping it, either by standing up and speaking out, or by joining ourselves together until our mass becomes critical and impossible to ignore. We can even do both – it's always lovely to see very vocal activists like designer Vivienne Westwood quietly adding her body to the throng of other bodies at a protest. Your ways of joining in can be as quiet or as loud as you like.

Books like this are fine. I promise I wouldn't be writing one if I didn't have the sincere idea, or wish, that it *might* help. But they can't – and shouldn't – pretend that there's a simple solution or a template *that will make everything OK*: "Do deep breathing." "Become self-sufficient." "Be an activist." "Go and see a shrink." People who offer, or accept, answers too readily are running away from difficult truths. Unfortunately, climate anxiety doesn't respond well to this kind of trickery – it's far too complex, and the people who suffer from it tend to be intelligent, critical thinkers; otherwise why would they be so afraid of something that takes an empathic, imaginative leap to understand at all? We have to be honest with ourselves and each other about the potential volatility of the situation we're in. Big, bad things really can happen, but we mustn't freeze our lives because of it. Not only may our worst nightmares turn out to be just that, but even catastrophe can contain its own kind of beauty – especially if you insist on finding it there. This can never have been truer than during the initial stages of the coronavirus pandemic, when not only did people exhibit true generosity towards one another, looking out for friends and neighbours, and showing passionate support for key workers, but in the extraordinary displays of nature – the clear water in the Venice canals, and the sudden appearance of friendly, playful dolphins, for example. If you can bear to let your anxiety be of the hopeful kind – which, in itself, is an extraordinary act of generosity – there's no need to give in to misery and helplessness. And the more you can cultivate this attitude in yourself, the more you will pass it on to others. Fake it so that we can *all* make it!

People, perhaps especially psychoanalysts (including me), often have a tendency to be pretty fatalistic about human nature. It's true that people are liable to be mean and naggy and selfish and prickly, but we also have it in us to be mega-amazingly generous and sweet. Lots of words have been spewed on the subject of whether we're fundamentally good or bad. Are we born innocent, after which "civilization" screws us up? Or are we, at heart, craven balls of greed and aggression, barely held in check by the laws and norms our cultures place on us? Of course, questions like this are unanswerable. Or people's attempts at answers are bound to be informed by their own experiences and prejudices. But perhaps there is a common experience we can all somehow relate to: the minute we are born, we land in the arms of other people. If we are still here, it's because those people held us – however well or badly. We must have reached out to those people for help. As Jonathan Lear very beautifully puts it:

> This is the archaic prototype of radical hope: in infancy we are reaching out for sustenance from a source of goodness even though we as yet lack the concepts with which to understand what we are reaching out for.[47]

Whether life then makes us shy or sociable, generous or mean, we all know what it is to depend on another for something; and if we have made it this far, it's because our hope wasn't altogether misguided. Even if our upbringings fell far short of the ideal, the fact that we kept going, kept asking, kept receiving, is perhaps proof enough that something in us wanted us to be here, and that something outside us did too.

If you're reading this, it's probably because you hope that other people's ideas can help you. And in a maternal way, I worry that nothing I offer can ever be enough. I can offer suggestions, present statistics, say what works for me. But ultimately you will make your own choices, find what works best for you, even if it's through a process of trial and error. I can't save you, and you probably can't save me – although I'll be waiting for your email now – but we can be tolerant of each other's shortcomings and share anything that seems to us worthwhile. This, to me, forms the basis of any solution to the crises that may come. We hold out hope for each other, that way exponentially increasing our chances.

HOPE: CHAPTER OVERVIEW

1. Greta Thunberg doesn't really mean it about the hope thing; she's surely one of the most hopeful people on the planet.
2. If you think you might have a good idea, don't be shy about it.
3. Anxiety + hope + courage + a capacity to tolerate uncertainty = a very strong foundation for facing the future.
4. By taking the climate crisis seriously, you are making yourself a part of something big.
5. You're born in a state of radical hope. Don't lose it!
6. Let other people hold and support you. I'm sure they will like it too.
7. Leave ruminating to cows.
8. I hope your mum is better at saying comforting things than I am. Still, I really do believe in the possibility of an interesting, worthwhile human future – and I very much hope that you do too. Let's try to make it happen.

10

RESILIENCE: BOUNCE BACK STRONGER

"Uncertainty is the very condition to impel man to unfold his powers."

Erich Fromm, psychologist and humanist philosopher

Of all the words and ideas that come up in relation to eco-anxiety, "resilience" is perhaps the most frequent, and possibly the most potent. This is what we will need to carry us through whatever's to come. Happily, the world and all its lifeforms have this quality in abundance. You could even say that life itself *is* resilience – a built-in, senseless force that pushes for its own continuation, just because it does. Life has tons of ways of coming back from the things that threaten it. Wounds heal, systems adapt, losses can be compensated. It isn't infallible – but the fact that it's fragile and limited is what makes life so gut-wrenchingly precious. Our existence on this orb is hard won. To have made the move from amoebas to molluscs to fish to mammals is outrageously, outlandishly tenacious. Life is brave – it'll try anything in order to self-sustain; it seems to keep choosing itself, as if it can't quite believe its luck: "Hang on in there, guys. This ride is good!" The idea that there are people out there who haven't quite clocked the unlikelihood of their – our – good fortune is peculiar. To put planetary life in jeopardy involves a blindness that's quite spectacular. How is it possible to be so unmoved by a miracle such as the one we all inhabit every day? This question is particularly perplexing as it's the safe, comfortable, rich people who are prepared to fry it for everyone. What's up with those guys? We need to speak to them urgently.

In this chapter we will look at what resilience might mean for human beings – what it is, and how we can encourage, develop and pass it on. We will also look at planetary resilience, and some of the amazing success stories taking place around us, demonstrating that all is far from lost. It seems that, alongside the crazy, wilful blindness, there's a massive, super-heroic push to right wrongs and get things back on track. I had no idea how hard people were trying until I Googled "environmental good news". Whatever I manage to jot down here will only scratch the surface – people are doing *incredible* things. Perhaps you're already one of them. Or maybe you will be. Anyhow, it would be nuts to succumb to pessimism with all this amazing stuff going on.

Individual Resilience

Resilience is a fairly broad idea, covering all sorts of qualities and approaches to life. At bottom, it's all to do with flexibility, and being able to recover from difficulties. This, of course, doesn't mean avoiding trouble, but being able to face it head on when necessary. Resilience involves being realistic about the problems you face, while still believing that life is worth living.

To go back to "anxiety", perhaps we could say that worrying too much about things that haven't happened yet is a bit of a double-edged sword. On the one hand, you might be suffering unnecessarily, but on the other, you may be actively engaged in useful activities that work against the threat. If you can somehow combine the second aspect of anxiety with the belief that you will handle whatever life throws at you in the best way you can, then you should be able to look the future in the face, even if it's scary.

The cornerstone of resilience is an ability to deal with uncertainty. It can be a difficult attitude to cultivate because the whole of modern life presents us with the illusion that uncertainty is avoidable. Rather than accepting it, we are encouraged to do everything we can to fight it.

Medicine makes illness seem like an unnecessary intrusion. Architecture protects us from the outside world. Economic systems pretend to self-regulate. Farmers employ technology to ensure a constant food supply. Our smartwatches order us to keep fit – our bodies must be kept in tip-top condition, even in old age. Even love can be made to look like a permanently available commodity – just keep swiping and we'll find it. It's as if any kind of deviation from absolute safety and certainty is an outrage, an anomaly, an unforgivable divergence from the plan. All this, of course, happens because life itself is inherently uncertain. You could get killed, you could get sick, you could starve. Our civilizations and technologies militate against this, inching ever forward in the war against the unknown.

However, it seems as though the fundamental law of uncertainty is way more powerful than anything we can throw at it; our comfortable, risk-averse culture is now the very thing that's threatening to destroy us. Our entitled overconsumption risks finishing us off. And still there are forces in society that try to convince us that the only thing we can do is to plough on regardless, putting our faith in the system to find ever more sophisticated means of holding contingencies at bay (although this attitude may become less tenable in the aftermath of the pandemic). No wonder anxiety is so prevalent. Rather than being allowed to worry realistically, we become alienated from our own fears. We are told it's irrational to fret; that so long as we conform to the ideals of our culture – work, save, exercise, shop, enjoy – we will be rewarded by a lifetime of perfect safety. Indeed, if we don't feel safe, it must be that we are getting something wrong. Either we are misunderstanding, we are failing or we are disobeying, and it's therefore our own fault if we don't feel happy and OK.

The responses to this are varied. Some might call for revolution. Some might sit back and wait for events to take their course. Some might run for the hills. Some might try to tweak their habits. No one can tell you which option to choose. Still, whichever you pick, you would be well advised to open your mind to uncertainty. A strong wish for certainty not only limits your options, but hugely contributes to anxiety. If life is only allowed to unfold along strictly proscribed lines, then it's bound to be haunted by the possibility of careering off-track. It's just not realistic to imagine that things will always go as expected. Earthquakes happen. Marriages end. Objects get lost. People die young. It's terrible, tragic, but it's also *real life*. While you may not want to invite tragedy in, you can also try to remember that bad, sad things are survivable. As long as you're alive, you can try to make the best of it, whatever 'it' is.

So, with uncertainty up there as the overarching principle, here's a list of other suggestions on how to make yourself the most resilient creature you can be.

1. Have feelings

Let yourself be as sad, angry, disappointed – or even as excitable and happy – as you need to be. Holding feelings at bay is exhausting, and can leave you feeling inauthentic and depleted. If the gap between your inner world and your outer presentation gets too big, it can become hard to function. It's also generally much easier to process feelings by expressing than by denying them. Feelings have an annoying habit of prodding until you give them some attention. If you attend to them quickly, they might even get off your case.

2. Enjoy nature

It's still there, *loads of it*, and it's the most impressive thing there is. Even the most advanced invention only exists thanks to plants and animals – none of us would be here without them. The fact that you are able to witness nature is surely enough to have made your life worth living, however things pan out. Just thinking about how it all got here can give you a mental orgasm!

3. Be grateful

Although life can be painful, it really is an extraordinary privilege to be here at all. Whom you direct your gratitude toward is up to you. It might be your god, it might be your parents, it might be your animal predecessors, or the strange act of chance that animated those first microorganisms. *Something* has put you here.

4. Be a good friend

Swapping ideas with people, knowing what goes on with them, and letting them know what goes on with you, makes life better. Sharing stuff, from thoughts to food to feelings to clothes, helps everyone. Try to find the best friends you can, though – the kindest, most thoughtful, engaged types who will return your trust and openness with interest. If there are none in your neighbourhood, look further afield, or online. Virtual communities can be great too.

5. Ask for help

And help others. As long as you do both, you'll know how rewarding it can be to be there for another person. It'll make you less afraid to ask for things when you need them. On the other hand, if you ask for help first then other people will know that they can ask back. In either case, it's generous.

6. Be accepting

While you might be consumed with fighting for change, it's also important to accept that some things are beyond your control. It can be hard to keep everything in perspective when the situation looks dire, but try to remember that there will always be good and bad things – and it's OK to enjoy the good things, even when the bad things are going full-throttle.

7. Be still sometimes

You can do it "formally" through mindfulness or meditation, or you can just stop doing stuff and notice how it feels to just *be*. It can be easy to get into living like a maniac, filling every minute with tasks, including the stressful task of sleeping for the correct number of hours. Stopping once in a while to notice the unlikely freakiness of just being here can be such a relief. You don't need to be useful all the time, nor happy. It's OK just to be.

8. Feel the floor

There is something underneath you that will stop you if you fall. Who could ever have come up with such a brilliant contraption? When things get too much, you can just drop to the floor and let it hold you. It never fails or lets you down. (Don't do it at the top of a steep hill, obvs, unless you're looking for excitement.)

9. Be self-aware

This can mean anything from knowing your strengths and limitations, to understanding your moods better. Therapy can be great for developing self-awareness, but so can honest conversations with friends and family. Self-awareness can help you to develop resilience by giving you a more flexible attitude toward yourself. If you're anxious or grumpy, maybe you have your reasons, but maybe *some* of your reasons are dodgy. Accepting that you're sometimes mistaken, flawed, and that your mind plays tricks on you, can help you to step down rather than get into unnecessary battles – including with yourself.

10. Befriend yourself

It's tempting to be much harder on yourself than you would be on someone else. (I know some people do the opposite, but I doubt they read ecological self-help books!) When you're giving yourself a hard time, stop and try to think what you'd say to a friend who was doing the same. I'm betting there's no way you'd be *that* mean to anyone other than *you*.

11. Take good news seriously

And seek it out. It can be easy to focus entirely on the things that are going wrong and to forget about the stuff that's going right. But good news is important because it demonstrates that there is a point in trying, and that it's possible to get good ideas off the ground.

12. Take breaks

It's really important to switch off. You might feel that constant activity is the only justifiable response to the crisis, but it's unsustainable. Focusing exclusively on the problem will definitely end up being bad for your health. People who work full-time on the climate crisis all report feeling stressed and needing to practise "functional denial" from time to time. It can be hard to switch off, so maybe you'll have to show some ingenuity in coming up with distractions that work for you. It could be sport, cinema, Bumble, Sudoku, cage fighting, sugar craft. I've had my ears pierced three times in the four months it's taken to write this book.

13. Be open-minded

Life could really be anything. The fact that you've been born into one particular set of circumstances doesn't mean you have to recreate them wholesale. It's OK to be poorer than your parents, to live differently from your school friends, or even to have no real idea of the life you want. Things could change drastically and still be OK.

14. Read and write

Exchanging ideas doesn't have to happen in person. There are so many brilliant books and articles about climate change: for example, Naomi Klein's *This Changes Everything*, Mike Berners-Lee's *There is No Planet B*, Greta Thunberg's *No One is Too Small to Make a Difference,* and Donna Orange's *Climate Crisis, Psychoanalysis, and Radical Ethics*. Although it can be alarming, it can also be comforting to see how other minds are coming at the problem. Producing your own writing too, in whatever form – novels, poetry, blogs, journalism, diaries – is a great way to spread ideas, or even just to externalize them. Things look different when you see them written down.

15. Be dumbstruck by the cosmos

If you think plants and animals are good, there's something even more astounding. It exists in a timescale that's almost painful to think about, and it operates according to beautiful, complex mathematical laws. The earth is only one tiny unimportant scrap of this huge interrelated system. However badly humans manage to fuck things up, we'll make absolutely no dent whatsoever in the properly grand scheme of things. In fact, the laws of the universe dictate that we're going to get fried anyhow. While this might not sound like the cheeriest news, it's certainly a reminder that we probably shouldn't get *too* hung up on "getting things right" on planet earth. Even if it all goes pear-shaped in our lifetimes, we'll have had our little window of opportunity to check in with the sublime.

Planetary Resilience, the Good News

We still have a few years to get things back on track down here on earth, so we might as well give it a go. Life *is* worth it, after all. And thankfully, some people are very much on the case. I hardly know where to start, because, when I think of the enormity of what they're doing, it makes typing almost impossible. Some of the most incredible work involves trees and reforestation. It's been estimated that it would take around a trillion more carbon-sucking, oxygen-spewing trees to get the atmosphere back on track. So finding the cheapest, fastest, most efficient planting methods is extremely important.

A tech company called BioCarbon Engineering have invented a drone that shoots pre-germinated seed pods into the ground. It can plant four times faster, and far more cheaply than human planters, as well as being better able to reach inaccessible land. The drones began by restoring mangrove trees in Myanmar, and have already planted over ten million between 2018 and 2019. Their initial work was such a success that new projects have taken off elsewhere. In 2019, a crowdfunded project in Canada, run by an organization called Flash Forest, raised $10,000 on Kickstarter, which will enable them to plant 150,000 trees by the end of 2020. They plan to have planted a billion trees by 2028, with the promise that the Canadian government will also plant at least another billion. They also aim to support biodiversity by planting at least eight different species of tree.

In other parts of the world there are record-breaking mass tree-plantings. In India in 2019, 220 million trees were planted in a single day. However, Ethiopia set a record by planting 350 million in one day, all of which is part of a massive initiative to plant 4 billion indigenous trees to replace those lost to drought.[48]

The way we produce energy is also changing. As of this decade, solar energy has become the world's preferred new source of power. Solar energy systems are being installed at a faster rate than those for any other type of energy, with wind coming second (and coal, unfortunately, still third). Together, solar and wind energy are now able to compete with coal on price, so there's no longer an incentive to stick with dirty fuels. Economists predict that this may bring about the rapid decline of the fossil fuel industry.[49]

Partly thanks to the shift of balance in the power industries, CO_2 emissions flat-lined in 2019, in spite of the global economy growing by just under 3 percent in the previous year. Some countries' emissions rose, while others fell, bringing about an unchanged total figure. The biggest fall in emissions for a single country was seen in America, which was down 140 million tons. The EU dropped by 160 million tons. And Japan by 45 million tons. Most of the increases came from countries in Asia, with their expanding coal-powered industries. However, with other developed countries demonstrating that industry can continue perfectly well with cleaner fuel sources, it seems the balance will continue to tip more and more toward solar and wind power across the globe.[50]

Humpback whales are back to 90 percent of their historic population of 27,500, after falling to a population of around 450 in the 1950s. This is thanks to restrictions on whaling put in place in the 1960s, and made more stringent in the 1980s.[51]

The National Pollinator Garden Network overshot its aim of a million pollinator gardens by 40,000 in 2019. Most of the gardens were registered in America, but the movement is growing and anyone with any outdoor space at all can join, as can public gardens. At last count, this added up to five million extra acres of pollinator-friendly planting. To be a part of this initiative, all you need to do is plant native plants and wild grasses, stop weeding so much, and refrain from using pesticides. To very slightly paraphrase the Bristol University Professor of Ecology, Jane Memmott: "[If y]ou can't personally help tigers, whales and elephants, you really can do something for the insects, birds, and plants that are local to you."[52] (I added the "if" at the beginning just in case you are in touch with tigers, whales and elephants. If so, may I visit you?)

The tiger population of India has grown 33 percent in just four years, from 2015 to 2019. Mountain gorillas (apparently the most protected species on the planet) are gradually increasing in number, thanks to intensive conservation efforts in Uganda. An extremely rare black leopard, not seen for 100 years, has recently been spotted in Kenya. Californian condors have come back from near-extinction. There were just 24 left in the 1980s, and now there are around 500. The world's largest breed of bee has reappeared in Indonesia, having apparently vanished for 38 years. A species of tortoise not seen since 1906 has been spotted in the Galapagos.[53]

Since 2012, the Zoological Society of London has teamed up with a company called Interface to reuse discarded fishing nets by turning them into carpet tiles. The nets are removed from the sea, saving fish from 'ghost fishing' (getting caught in unused nets), while the conversion of nets into carpets provides an income to people in small coastal communities. By 2016, 137 tonnes of plastic fishing nets had been pulled out of the sea, and 900 families had received money from the initiative.[54]

Perhaps the best news of all is that *the world has finally caught onto the fact that the climate emergency is real.* The cat is well and truly out of the bag. Businesses and governments are already scared of people power and are feeling the pressure to rethink their strategies accordingly. Supermarkets stock delicious vegan food. Insurers are afraid of being seen to invest in fossil fuels. More than 1,200 local authorities in Britain declared a climate emergency in 2019, and numbers are likely to increase massively in the coming years. Even the British Conservative government is getting behind environmental issues, speeding up the changeover to electric cars, supporting cycling initiatives and investing in replacing old-fashioned gas boilers and insulating homes. And if it turns out that they're only making little gestures to keep us quiet (who, the Tories?), we now have a powerful enough voice to answer back. The USA's carbon output is falling, in spite of their withdrawal from the Paris Climate Agreement. Seventy-five percent of Americans now agree that climate change is caused by humans, in spite of massively funded campaigns to persuade them otherwise. Countries like Costa Rica are proving to the world that it's possible to make huge changes in a short space of time. In 2018 their economy grew by 3 percent, and 98 percent of their electricity came from renewable sources. They are well on track to be 100 percent carbon neutral by 2050.[55]

It seems that we are winning the battle to get people to recognize that it's time to act. Our anxieties are paying off. If you're worried about the climate, thank you. You are one of the people who's made all this happen. Without your anxiety, the situation would be far worse. So let's stick together, keep going, and try to enjoy the fight for life as much as we can.

OVERVIEW (FOR THE WHOLE BOOK THIS TIME ...)

1. Understand that your anxiety is a brilliant adaptation. Your body is genius: it's telling you to try to save our excellent planet. Join up with other worried people and not only will you feel relieved, but you will be helping to create a superpower. We're winning!
2. Don't get lost in the future. Stay here, because here is where we can get stuff done.
3. Look after yourself, others and the planet. All three, equally. You won't be able to do the last two if you don't do the first one. It's not selfish, it's kind.
4. Become the best conversationalist you can. Talk to people. Listen to them. Be considerate and honest, not to mention subtly persuasive when necessary. Good communication isn't about showing off or impressing people; it's about being there with others in ways you can all enjoy and learn from.
5. Stay open to the world and all its possibilities. Know that you will do the right thing by other people in an emergency.
6. Let children say whatever they need to about the climate crisis. You can be reassuring without being dishonest. And have a baby if you want. Who knows what's going to happen? It would be so annoying to hit menopause and realize that it would actually have been OK.
7. Appreciate the amazing things this planet has to offer, and don't worry if some of these are ecologically suboptimal. It's not your fault if you've been brought up to like the odd polluting activity. It might take time to adjust to a different set of ideals.

8. Combine environmentally responsible personal choices with activism and social engagement. It's not one or the other; it's both.
9. Proceed as though it's possible to make a difference. Celebrate and share your eco-wins, however small.
10. When terrestrial life gets too much for you, let your mind drift up to the stars. It's so easy to forget they're there …

TALKING TO MY THERAPIST ABOUT CLIMATE ANXIETY

So we sit, as we do
every Tuesday, in chairs that are somehow
too deep, with the 6 feet
of professional distance
spread out on the rug before us.

How was your week?

Not great – I wring my hands – not great
I am not working as hard as I could be and my sister won't talk to me
and my mum has a cold and I'm terrified she'll die
and I can't sleep because I'm up at night tumbling into terror about our approaching climate catastrophe.

Usually my therapist replies by listening quietly,
watching – not
saying much
until the twenty minute deconstruction of my suffering at the end of the hour,
and sometimes
if I am crying,
she will tell me that
nobody is going to die.
This time, she simply nods.

Nadia Lines (2019)

NOTES

1 Responsible news outlets are changing the language they use to match the seriousness of the situation: "global warming" has become known as "global heating", while "climate change" is now apparently better conveyed by the terms such as climate "emergency", "crisis" or "breakdown".

2 Philosophy Tube: https://www.youtube.com/watch?v=CqCx9xU_-Fw

3 Jem Bendell, "Deep Adaptation: A Map For Navigating Climate Tragedy" (2018), available at: https://jembendell.com/2019/05/15/deep-adaptation-versions/

4 Joseph Dodds, *Psychoanalysis and Ecology at the Edge of Chaos: Complexity Theory, Deleuze/Guattari and Psychoanalysis for a Climate in Crisis* (Routledge, 2011), p.73.

5 Of course, there are people who would disagree entirely. There is another strand of thinking – "green capitalism" – which suggests that science will solve all our problems for us. The argument goes that within the next few years, the excess carbon will be sucked out of the atmosphere, meat will be produced synthetically, plastic will be munched by harmless bacteria, and the whole world will heal itself without us having to lift a finger, so keep shopping! More on this in the next chapter…

6 Jem Bendell, "Deep Adaptation".

7 Harold Searles, "Unconscious Processes in Relation to the Environmental Crisis" in *Psychoanalytic Review* (1972), p.361.

8 Ibid., p.366.

9 Ibid., p.367.

10 See report and related research at: https://www.ncbi.nlm.nih.gov/pmc/articles/PMC6020909/

11 Madeleine Stone, "How Antarctica is Melting from Above and Below" in *National Geographic*, October 2019.

12 Kelton Minor *et al.*, "Greenlandic Perspectives on Climate Change 2018–2019: Results from a National Survey", available on ResearchGate: https://www.researchgate.net/publication/339177908_Greenlandic_Perspectives_on_Climate_Change_2018-2019_Results_from_a_National_Survey
13 https://www.bloomberg.com/news/articles/2017-12-20/are-miami-beach-s-luxury-towers-the-future-of-climate-resilience
14 Janis L. Dickinson, "Why Climate Change Threatens Our Inner Life and Survival" on *Cornell Chronicle* website (21 January 2010). See: https://news.cornell.edu/stories/2010/01/climate-change-threatens-our-inner-and-outer-lives
15 J. L. Dickinson, "The People Paradox: Self-Esteem Striving, Immortality Ideologies, and Human Response to Climate Change" in *Ecology and Society* (2009), 14(1): 34. Available online at http://www.ecologyandsociety.org/vol14/iss1/art34/
16 José M. López, "Profiles in Vengeance: The Quest for a Chicano Gang Worldview" in *Mexican Studies/Estudios Mexicanos* (Summer, 1991), 7(2), pp. 319–329.
17 J. L. Dickinson, "The People Paradox".
18 Sally Weintrobe, *Engaging with Climate Change* (Routledge, 2012), p.205.
19 See US Department of Health and Social Services Help and Resources, National Opioids Crisis: https://www.hhs.gov/opioids/about-the-epidemic/index.html
20 Read Montague *et al.*, "Nonpolitical Images Evoke Neural Predictors of Political Ideology" in *Current Biology* (2014).
21 Robert Waugh, "Right-wingers Are Less Intelligent Than Left-Wingers, Says Study" in *Daily Mail* (1 February 2016).
22 Quoted in Owen Bowcott, "Lady Hale Warns UK Not to Select Judges on Basis of Political Views" in *Guardian* (18 December 2019).
23 George Monbiot, "The Problem is Capitalism" in *Guardian* (30 April 2019). Available at: www.monbiot.com/2019/04/30/the-problem-is-capitalism/
24 Extinction Rebellion, *This is Not a Drill: An Extinction Rebellion Handbook* (Penguin, 2019), p.99.

25 Walter Benjamin, *Illuminations* (Fontana Press, 1992), p.101.
26 Nick Tilsen, "Building Resilient Communities: A Moral Responsibility", TEDxRapidCity (14 July 2015). Available at: www.youtube.com/watch?v=e2Re-KrQNa4
27 Ibid.
28 Douglas Murray, "Terrifying Our Children With Doom Mongering Propaganda on Climate Change is Nothing Less Than Abuse" in *Mail on Sunday* (19 January 2020).
29 Christopher Hooton, "Sir David Attenborough: Humans May Be an Endangered Species" in *Independent* (30 December 2014).
30 Trent MacNamara, "Liberal Societies Have Dangerously Low Birth Rates" in *The Atlantic* (26 March 2019).
31 https://www.theguardian.com/environment/2016/feb/17/how-green-is-online-shopping
32 See: www.theguardian.com/membership/video/2014/oct/29/vivienne-westwood-capitalism-clothing-video
33 Ashitha Nagesh, "Investigation Launched into Chef Who 'Spiked Vegan Group's Meal With Meat'" in *Metro* (2 Jan 2018).
34 Kimberly Amadeo, "Why Trickle-Down Economics Works in Theory But Not in Fact" in *The Balance* (27 October 2019). Available at : www.thebalance.com/trickle-down-economics-theory-effect-does-it-work-3305572
35 Malcolm A. Weiss, John B. Heywood, Elisabeth M. Drake, Andreas Schafer, and Felix F. AuYeung, "On the Road in 2020: A Life-Cycle Analysis of New Automobile Technologies" (MIT Energy Laboratory, 2000). Available at http://web.mit.edu/energylab/www/
36 Available at: www.greencarreports.com/news/1093657_buying-a-new-car-is-greener-than-driving-an-old-one-really.
37 Figures taken from Seth Wynes and Kimberly A. Nicholas, "The Climate Mitigation Gap: Education and Government Recommendations Miss the Most Effective Individual Actions" in *Environmental Research Letters* (Institute of Physics, 2017), 12(7).
38 M. Springmann, M. Clark, D. Mason-D'Croz, *et al.*, "Options For Keeping the Food System Within Environmental Limits" in *Nature* (2018), 562, pp.519–525. Available at https://doi.org/10.1038/s41586-018-0594-0

39 "New study: Vegan diet reduces carbon footprint by 73%" in *Vegconomist* (19 July 2019).

40 See: https://wrap.org.uk/sites/files/wrap/Food_%20surplus_and_waste_in_the_UK_key_facts_Jan_2020.pdf

41 LifeGate is an Italian organization that's been running since 2000. It describes itself as a "hub of sustainable innovation". For more information, visit: https://www.lifegate.com/people/lifestyle/city-trees-smog-pollution

42 Statistics released by the United States Environmental Protection Agency. See: www.epa.gov/greenvehicles/greenhouse-gas-emissions-typical-passenger-vehicle

43 Available at: www.dazeddigital.com/politics/article/46460/1/extinction-rebellion-climate-change-activism-middle-class-issue

44 Jonathan Lear, *Radical Hope: Ethics in the face of cultural devastation* (Harvard University Press, 2008), p.105.

45 "Why the Psychology of Climate Change?" in *The Psychotherapist*, guest-edited by Judith Anderson and Chris Robertson (Summer 2016), Issue 63, p.6.

46 Lear, *Radical Hope*, p.123.

47 Lear, *Radical Hope*, p.122.

48 www.onetreeplanted.org

49 www.happyeconews.com

50 www.goodnewsnetwork.org

51 www.earthsky.org

52 www.goodnewsnetwork.org

53 https://onetreeplanted.org/blogs/stories/good-news-july and https://medium.com/future-crunch/99-good-news-stories-you-probably-didnt-hear-about-in-2018-cc3c65f8ebd0

54 www.interface.com

55 www.theconversation.com

FURTHER READING

Ernest Becker, *The Denial of Death* (Souvenir Press, 2011).
Mike Berners-Lee, *There is No Planet B: A Handbook For the Make or Break Years* (Cambridge University Press, 2019).
Rachel Carson, *Silent Spring* (Penguin Classics, 2000).
Lily Cole, *Who Cares Wins: Reasons For Optimism in Our Changing World* (Penguin Life, 2020).
Extinction Rebellion, *This is Not a Drill: An Extinction Rebellion Handbook* (Penguin, 2019).
Naomi Klein, *This Changes Everything: Capitalism vs the Climate* (Penguin, 2015).
Bessel van der Kolk, *The Body Keeps the Score* (Penguin, 2014).
Jonathan Lear, *Radical Hope: Ethics in the Face of Cultural Devastation* (Harvard University Press, 2008).
Karen Logan, *Clean House, Clean Planet: Clean Your House For Pennies a Day, the Safe, Nontoxic Way* (Simon & Schuster, 1997).
Donna M. Orange, *Climate Crisis, Psychoanalysis, and Radical Ethics* (Routledge, 2016).
Greta Thunberg, *No One is Too Small to Make a Difference* (Penguin, 2019).
Sally Weintrobe, *Engaging with Climate Change* (Routledge, 2012).

RESOURCES

(The following listings are for information only and do not imply any endorsement of the arguments in this book.)

Australian Student Environment Network
Connects student environment groups from around Australia committed to building grassroots movements for change.
https://asen.org.au/

Climate and Mind
Aims to explore how climate change impacts our thoughts, emotions and behaviour, bringing together resources and ideas from a range of disciplines.
www.climateandmind.org

Extinction Rebellion
An international movement that uses non-violent civil disobedience in an attempt to halt mass extinction and minimise the risk of social collapse.
https://rebellion.earth/

Greenpeace
An environmental organization that expose global environmental problems, and which aims to force solutions for a green and peaceful future.
https://www.greenpeace.org/international/

LifeGate
An Italian organization that describes itself as a “hub of sustainable innovation”. https://www.lifegate.com/people/lifestyle/city-trees-smog-pollution

One Tree Planted
An environmental charity focused on global reforestation.
www.onetreeplanted.org

Peoples Climate Movement
US-based organization that uses two key strategies to demand bold action on climate change: mass mobilization and movement alignment. peoplesclimate.org

Project Greenworld International
The world's biggest children's environmental initiative, with groups in Oman, India, Bahrain, Korea, Ghana, Nigeria, Nepal, Sierra Leone, Uganda and Ethiopia.
projectgreenworld.wixsite.com/projectgreenworld

UK Student Climate Network (UKSCN)
A group of mostly under-18s taking to the streets to protest the government's lack of action on the climate crisis in support of YouthStrike4Climate. https://ukscn.org/

Wretched of the Earth
A collective of over a dozen grassroots indigenous, black, brown and diaspora groups, individuals and allies acting in solidarity with oppressed communities in the Global South and Indigenous North.
https://wretched-of-the-earth.tumblr.com

ACKNOWLEDGEMENTS

Primarily, thanks to Anya Hayes for commissioning this book in the first place. If you hadn't asked I certainly wouldn't have thought of it. And to Toby Moses at the *Guardian*, without whom Anya would never have asked. Also thanks to Ed Gillespie for taking the time to read the book and to write such a powerful foreword, and to Sue Lascelles for saving me with your kind and careful editing. To Jayoon Choi for your cover illustration — and for being such a good party guest. And, on that subject, thanks to *everyone* who came to my 50th. It really cheered me up in the middle of this depressing, lonely research. Oh, and to Devorah Baum and Darian Leader for your *excruciating* speeches. Thanks to Susan Morris and Benjamin Spiers for being my surrogate shrinks. To Sigmund Freud and Jane Austen; can't imagine life without you. To my mum and dad for being so understanding about the occasional perimenopausal outburst — and to my dad for mentioning The Hollies at a key moment. Also to Josh, Manny and Isiah Appignanesi, Jacqueline Lenoir, Cedar Lewisohn, Patty Ellis, Naomi Sylvester, Aruna and Luke Szarowicz, Jimmy and Meg, Nicholas Blincoe, Laura Morris, Andrea Arnold, Hannah Mumby, Nathalie Olah, Alison Ryder-Cook, Dr Clare Smith, Nadia Lines, Naja Marie Aidt, Jack Webb, Annelise Howard-Phillips and family, Cosmo Landesman, and above all to Dot for surpassing every single hope and expectation I had around being a mother. I'd never have dared to imagine I'd get to meet YOU!

INDEX